INFORMATION VISUALISATION

Sara Miller McCune founded SAGE Publishing in 1965 to support the dissemination of usable knowledge and educate a global community. SAGE publishes more than 1000 journals and over 800 new books each year, spanning a wide range of subject areas. Our growing selection of library products includes archives, data, case studies and video. SAGE remains majority owned by our founder and after her lifetime will become owned by a charitable trust that secures the company's continued independence.

Los Angeles | London | New Delhi | Singapore | Washington DC | Melbourne

INFORMATION VISUALISATION

FROM THEORY, TO RESEARCH, TO PRACTICE... AND BACK

Maria dos Santos Lonsdale

Los Angeles | London | New Delhi
Singapore | Washington DC | Melbourne

Los Angeles | London | New Delhi
Singapore | Washington DC | Melbourne

SAGE Publications Ltd
1 Oliver's Yard
55 City Road
London EC1Y 1SP

SAGE Publications Inc.
2455 Teller Road
Thousand Oaks, California 91320

SAGE Publications India Pvt Ltd
B 1/I 1 Mohan Cooperative Industrial Area
Mathura Road
New Delhi 110 044

SAGE Publications Asia-Pacific Pte Ltd
3 Church Street
#10-04 Samsung Hub
Singapore 049483

Editor: Umeeka Raichura
Assistant editor: Hannah Cavender-Deere
Production editor: Imogen Roome
Copyeditor: Sarah Bury
Marketing manager: Ben Sherwood
Book design: Maria dos Santos Lonsdale,
           Matthew Baxter and Wei Wu
Additional cover design: Shaun Mercier
Typeset by: C&M Digitals (P) Ltd, Chennai, India

Library of Congress Control Number: 2022938542

British Library Cataloguing in Publication data

A catalogue record for this book is available from
the British Library

ISBN 978-1-5297-2580-3
ISBN 978-1-5297-2579-7 (pbk)

To my loving children, Nuno and Rafael, and my husband David.

CONTENTS

INTRODUCTION

01

UNDERLYING THEORIES FOR INFORMATION VISUALISATION

02

FROM THEORY, TO RESEARCH, TO PRACTICE

03 FROM STATIC, TO INTERACTIVE, TO MOTION

3.1 Interactive Infographics

3.2 Motion Graphics

3.3 Intermotion Graphics

04

USER-CENTRED RESEARCH METHODS FOR VISUALISATION

4.1 Discovery Stage

05 FROM THEORY TO RESEARCH, TO PRACTICE... AND BACK

5.4 Case Study 4 | Surgery Recovery

5.5 Case Study 5 | Chinese Characters

5.6 Case Study 6 | Obesity Prevention

ACKNOWLEDGEMENTS

I would like to start by thanking all the staff and students with whom I have had the honour to work with at the School of Design and other Schools in the University of Leeds (UK) from 2015. The work we produced together, or that was supported by them, served as the inspiration for this book.

My gratitude also goes to the many people who took part in my research and helped me to understand better how information design can be such a powerful tool for improving our lives and wellbeing.

I would like to thank two of my former MA and PhD students in particular. Dr Matthew Baxter and Mr Wei Wu, for helping me design this book and the visualisations developed to accompany the text.

A heartfelt thank you also goes to Rune Pettersson, my biggest inspiration in the field of information design, for writing the foreword for this book. It is a true honour.

Special thanks go to the reviewers who gave excellent feedback to improve the book further, as well as to the editors and all at Sage who helped make my first book a true joy to work on.

Thank you as well to *Information Design Journal and Visible Language* journal, for allowing me to refer to the papers I published with them in Chapter 5 as case studies.

Finally, and the biggest thank you of all, goes to my beloved family who are everything to me and who inspire me to do my best every day.

ABOUT THE AUTHOR

In 1996 Maria received a four-year degree (Licenciatura) in Visual Communication from ESAD – Escola Superior de Art e Design, Matosinhos, Portugal. In 2006 she received her PhD from the Department of Typography and Graphic Communication, University of Reading, UK.

Maria dos Santos Lonsdale is a Professor of Information and Communication Design at the University of Leeds, UK. She has experience of teaching design theory and practice, and research methods for design in higher education, both in Portugal and in the UK. She has also worked as a design practitioner since 1996, having established her own design studio (in partnership) at the time.

Her main areas of research are: Information Design, Communication Design and Typographic Design. Her research is notable in the field of Design for its interdisciplinary nature, with collaboration across sectors such as Healthcare, National and Global Security, and Education. Her research is also notable in the field of Design for its research-based and human/user-centred approach, using a wide range of innovative research methods. At its core, her research focuses on the effective use of design to improve user performance and wellbeing.

She is a member of the International Society of Typographic Designers, and an Academic Advisor and Information and Communication Design Lead for various organisations, public services and companies in the healthcare sector. She is the Editor-in-Chief for the *Information Design Journal* and was previously Associate Editor of *Visible Language*.

FOREWORD

I have had the privilege of reading 15 publications by Professor dos Santos Lonsdale. Therefore, my expectations were very high when I got the opportunity to read this book before printing. Through her texts and my own work, we got to know each other.

Many 'communication related problems' started in the mid-1990s, when 'a total lack of design' gradually became an inherent aspect of the World Wide Web. Suddenly, almost anyone with easy access to a desktop computer and a few software programs could produce, publish, and widely distribute their own texts and their own pictures.

During the late 1980s and the 1990s I listened to several speakers at media conferences who said 'anyone–anywhere–anytime will be able to present their own ideas digitally'. Soon the WWW was more or less 'invaded and infected' by millions of amateurish, homemade personal documents, regarding different subjects, posted/published on different 'web-pages'. The result was often chaotic verbal and visual environments. Apparently, the concept 'anyone–anywhere–anytime' did not mean that everyone could become a skilled editor and, at the same time, a skilled graphic designer. As a consequence, we were faced with many examples of poor legibility as well as poor readability.

It is really not at all enough to have access to good equipment. This problematic issue has continued to escalate. It seems that members of the general public far too often are innocent 'victims' here. Sooner or later, we will all be exposed to information materials that not only are boring and unattractive; sometimes it is more or less impossible to read the verbal and visual messages, and to be able to understand the 'intended contents'. In the worst case, it may lead to serious misunderstandings, for example in matters relating to health and wellbeing.

If you want to publish some of your ideas and information on the WWW, or publish it as traditionally printed documents, it is always a good idea to first make sure you acquire some basic knowledge about communication. You really need to get a clear understanding about our human cognitive conditions. Thankfully, this is much easier now! In this new book Professor dos Santos Lonsdale demonstrates how, why, and what we need to do, and what to avoid.

In her introduction to this important book, Professor dos Santos Lonsdale writes: 'In an age where communication of information is faster than ever due to quick advances in technology and the ability to be always on(line), good information design is also more important than ever. Information visualisation in particular has the power to take large chunks of information and present these in a visual, concise and accessible way, consequently reducing information load.' I totally agree with this. Information visualisation is an important part of information design. Information visualisation includes analysis, planning, presentation and understanding of a message, its content, language and form. The main objective is to provide the information needed by the intended receivers/users in order to perform specific tasks. Information visualisation is a *process* (verb) as well as a *result* (noun) of that process.

When words and pictures are produced for informative purposes, it is always a good idea to start by trying to 'visualise' the information to be conveyed to the intended readers/users. Visualising a message means that we attempt to materialise it in an effective synthesis. Visualisation is usually a complex task, never a single act on its own, and it requires the close collaboration of several different experts.

In her introduction to this book, Professor dos Santos Lonsdale notes that many visualisations of information are poorly designed: 'They are cluttered and disorganised. They randomly use visual elements for decoration instead of function. They disregard legibility principles (especially in terms of text and colour). They are difficult to interpret and understand and end up putting the user off from wanting to engage with the information.' I totally agree with these conclusions, and I have seen this increase over many decades. In my own work as designer, editor, researcher, and teacher, I, and my co-workers, have always wanted to put *humans at the centre* (children as well as grown-ups), *never the technology*. When introducing computers in schools, in the early and mid-1980s, for example, this should mean that all students get better access to information and the process of learning becomes easier. A common basic attitude from our team of researchers in this area is: 'Computers and software must be adapted to the students' conditions. It is not right to force people to adapt to technology.' However, this has become a major problem in our modern societies.

A traditional academic discipline, one *field of study*, is a *branch of theoretical knowledge* that is researched and taught in higher education. A field is ultimately defined at least in part by its research questions. The boundaries of a discipline mark what falls *within* its breadth, and also what it *excludes*. In theoretical disciplines, students need to develop their theoretical skills. They need to work with theoretical assignments and exercise their analytical and logical skills. In established disciplines, research is often based on one of many well-known theories. Researchers formulate new hypotheses and follow established and reliable processes for research.

Most academic disciplines of today have their roots in the mid to late nineteenth century. In the early twentieth century, new important disciplines, such as *education* and *psychology*, were added. Many new disciplines focusing on specific, and sometimes narrow, themes were added in the 1970s and 1980s.

Information Design is a very young academic discipline, but not at all a new area of knowledge. Humans have sporadically occupied the Blombos Cave in South Africa. As early as 100,000 years ago some individuals produced engraved pieces of ochre, with abstract geometric designs. Today, no one understands the meaning with these designs. However, they are regarded as the oldest known preserved human 'artwork'.

In Egypt, people produced and sold the so-called '*Books of the Dead*' at least 4000 years ago. These hand-made scrolls of papyrus contained an *integrated design* with a verbal and visual amalgam with advice about the necessary trip to the Kingdom of Death. The *Books of the Dead* may be the first examples ever where text, pictures and graphic design really are fully integrated.

Our discipline, Information Design, was never 'divided away' or 'branched from' another, already well-established discipline. Rather, Information Design was deliberately 'put together' with different important elements from *several sources* of experience, and solid practical experience. Basically, this happened at the same time, in different parts of the world, in the late 1990s.

The discipline Information Design has a practical as well as a theoretical part. The same is true for architecture, dance, design, economics, education, engineering, fine arts, journalism, medicine, music, and even more disciplines. In all these *fields of knowledge* it is a major and difficult, but necessary, challenge to find a good balance between *practical experience* and *theoretical knowledge*. The study of information design is a broad area with contacts to several other areas of practice and research, very much like the ongoing developments in architecture and building construction, and of engineering and technology. In an *applied science*, people apply basic existing scientific knowledge to develop new practical applications for different needs.

Academic research in Information Design has a pragmatic perspective, on different kinds of knowledge. It is obvious that each 'research problem' needs its own, specific research method. There is never only one research method in Information Design. New findings are tested, and the results are confirmed in different environments, and in different situations. In order to describe research in Information Design we may use words like *creativity*, *flexibility*, and *practical testing* in both experimental and in real-life settings. Working with research in Information Design is a challenging occupation. This is partly due to this complexity.

We may view Information Design as an 'applied science', as a 'combined discipline', as a 'practical theory', or as a 'theoretical practice'. I totally agree with Professor dos Santos Lonsdale when she writes the following in the *Introduction* to this book: 'the optimum information design is achieved by a triangulation between theory, research and practice. Each of these three components should have equal weight, work in synchrony and as a continuum. This is the ethos of this book that will take you on a journey from theory to research, to practice... and back.'

Personally, and in line with Professor dos Santos Lonsdale's approach to 'design with the user and for the user', I would also include the intended audience, the readers, the users, in this model and change the one flat equilateral triangle into a *regular equilateral pyramid*, where all four sides are equilateral triangles. In this model, *practice*, *research*, *theory*, and *users* each occupy one flat equilateral triangle.

In Information Design the *main goal* is *clarity of communication*. Here, *clarity* refers to 'the quality of being clear and easy to understand', and to 'the ability to think clearly and not be confused'. A third meaning of 'clarity' is 'the quality of being easy to see or hear'. The expression *clarity of communication* refers to the two concepts *legibility* and *readability*.

This book typifies this clarity of communication as the main goal of Information Design through its content and also in the way it utilises visualisation to communicate in a clear, effective, and engaging way.

Rune Pettersson

PhD, Retired Professor of Information Design at Mälardalen University in Sweden.

0

Introduction

THE CONTEXT FOR THIS BOOK

In an age where communication of information is faster than ever due to quick advances in technology and the ability to be always on(line), good information design is also more important than ever. Information visualisation in particular has the power to take large chunks of information and present these in a visual, concise and accessible way, consequently reducing information load.

Historically, information visualisation was often associated with data and the visualisation of patterns and trends in abstract datasets. In more recent and advanced research, and for the purpose of this book, information visualisation encompasses both the visualisation of information through graphics (i.e., infographics – static, motion or interactive) and the visualisation of data to better understand it. In addition, information graphics (infographics) are considered in this book as a branch of information visualisation. They represent the visualisation of information combining graphics, text, colour, layout and data. Data visualisation is, in turn, considered in this book as a branch of infographics and one of the design elements that is often included in infographics. The focus will be on the most established and researched forms of data visualisation. However, the same general principles will apply to other types of data visualisation.

Despite the power of information visualisation to communicate information in a digested, clear, efficient and engaging manner, more often than not this is not achieved by infographics or data visualisation in the public domain. Many visualisations of information are poorly designed:

- They are cluttered and disorganised;
- They randomly use visual elements for decoration instead of function;
- They disregard legibility principles (especially in terms of text and colour);
- They are difficult to interpret and understand and end up putting the user off from wanting to engage with the information.

In some cases, they actually provide inaccurate information and are misleading, creating confusion and doing more harm than good. How often do we see, for example, bubble charts in infographics that are used simply for aesthetic reasons? But when we look closely, the size of the bubbles does not reflect the data they should represent and leads us into error. Not to mention that these are quite complex to understand by the layperson and any design solutions must consider the target audience. This is particularly worrying when it happens in areas where inaccurate information can have dramatic consequences, such as healthcare and security.

The inaccurate perception that information visualisation is primarily aesthetics and only then function, is equally worrying. How often do we see infographics and data visualisation looking more like a catalogue for all that is possible to do with software and information technology? For example, the use of several colours, bright colours that vibrate when close together, colour backgrounds, shade, 3D charts, pie charts, donut charts, bubble charts, exploded charts, 3D text, outline text, colour text, small text, large amounts of text in all-capitals... How overwhelming is all this just by reading it? How much worse will it be to look at it and try to process and understand the information at the same time?

Naturally, it is important to grab the user's attention and appeal is intrinsic to well-visualised and organised information. Infographics do not need to look boring and simplistic to be functional. Both function and aesthetic should be present, but in equilibrium and with meaning and purpose.

Although all this is unfortunate and disappointing from an information design's perspective, it is not an exaggeration, and it is not surprising either. Communicating information in a digested, clear, accessible, efficient and attractive way requires knowledge and expertise, mastery and skill, time and commitment. While this book cannot contribute or control how much time and commitment one dedicates to visualising information, it can certainly increase one's knowledge, expertise and mastery of information visualisation.

WHY THE NEED FOR THIS BOOK?

Based on my experience as a design practitioner, educator, academic and researcher, the optimum information design is achieved by a triangulation between theory, research and practice. Each of these three components should have equal weight, work in synchrony and as a continuum. This is the ethos of this book that will take you on a journey from theory to research, to practice... and back.

Contrary to what most of my students believe when they start their Bachelor or Master's degree, practice, while the main drive for a design practitioner, is not more important than theory or research. Practice is simply the materialisation of knowledge from theory and research findings at the stage of development during the design process. This is what is missing for some designers and why they decided to pursue a PhD. They want to understand, to explore first-hand and to find the answers to the many whys, the hows and the buts they have as a practitioner. While not everyone can or should have to do a PhD, we can certainly benefit from our peers' knowledge, as is the case in this book.

There are two very simple but genuine and important things that I have learned throughout my academic journey, which I generously share with my students and colleagues. To be a good information and communication designer we need to be patient and pay attention to detail. Patience is required because it is important to spend a considerable amount of time:

- Finding out what other authors and researchers have done and found;
- Developing, testing and iterating our design solutions;
- Generating our own findings (as a practitioner and/or researcher) that in turn will validate our design solutions.

Acquiring empirical knowledge and making informed decisions based on this knowledge is the greatest advantage an information designer can benefit from. We should be creative and strategic thinkers, as well as problem solvers. These soft skills are what distinguishes our trade as designers from many other trades, and what makes us bullet-proof to automation, which is becoming more and more prevalent with the fast advance of technology.

Patience also means that before starting to create and design our solutions, we must understand our target audience very well: their needs, expectations, demographic, and any other characteristics of relevance to the design solution that should be tailored for them. Skipping or rushing this stage means failing as a designer because the solutions are more about our preferences as a designer than about the needs of the audience that we should help and accommodate.

The combination of patience and attention to detail makes our information design solutions not only effective, successful and meaningful, but also directly applicable to real-life contexts. This includes looking at every detail with caution and placing every piece of information on the page/platform in a purposeful and strategic manner. The most efficient way to achieve this is to also involve the users and stakeholders during the problem identification and ideation stage (that precedes the design development stage) through a co-design approach.

This attention to detail is therefore also closely related to the user and their involvement in the design process. Despite our excellent knowledge as designers, our tacit and intuitive decision making, our ability to think strategically and creatively, we are unable to create the optimum design solution unless we know exactly why and for whom we are designing. This means that the user must continue to be involved in the design process until the last stage of implementation, during which we should ascertain one last time that the design is fit for purpose. In sum, we should design with the user and for the user if we are serious about improving people's quality of life and facilitating their interaction with information at a time of information overload. This is exactly how this book was developed and designed, i.e., patiently, paying attention to detail and developing it with the user and for the user.

This book is therefore, in itself, an example of an information design output that sits on months of research, various stages of design development, testing, iteration and validation. A mixed-methods and human-centred design research and testing approach was used, and cognitive theory and design principles were applied to fully meet users' needs and expectations. All in all, the objective was to find a design solution for this book that allows you, the reader, independently of your background (design practitioner, student, educator, academic and/or researcher) to process and understand information visualisation guidance as clearly as possible and then apply it to practice in an effective and efficient manner.

In sum, it is by building on these foundations that this book can help those interested in information design to develop their knowledge and skills further. This triangulation and equilibrium between theory, research and practice, with the user centre stage, is the solid skeleton of this book. The illustrations, the practical examples and case studies showing direct application to real-life contexts, all created and/or tailored and tested exclusively for this book, are the flesh of this book and the visual materialisation of its content. A book of this nature is highly needed and long due in the field of
Information Design.

HOW THIS BOOK CAN HELP

This book will take you on a staged and sequential journey by:

1. Explaining and discussing information visualisation from both a scientific and practical perspective.
2. Gathering findings and producing guidelines that bring research and practice together and that take visual perception and cognition into account.
3. Delivering those guidelines through information visualisation itself, i.e., doing what I advise in this book.
4. Offering real examples of application to practice and clearly identifying the dos and don'ts of different approaches.
5. Focusing on research methods that equip creators of information visualisation with the necessary tools to design with and for the end user, i.e., to understand their audience at the initial stage of the design process, to iterate and refine their concepts with the involvement of the user, and to validate the design solutions in
    real-life contexts.
6. Presenting case studies that consolidate all of the above through research studies that are user-centred and have a direct application to practice, but that extend beyond business and journalism, which are the main areas where information visualisation has been explored thus far.

CHAPTER 1 | UNDERLYING THEORIES OF INFORMATION VISUALISATION

The rule of thumb in this book is always to find a balance between what is primarily efficient and clear information visualisation and what is ultimately also aesthetically pleasant and engaging. To be able to achieve this balance and assure that information visualisations are fit for its users and well designed, information theories from cognition (cognitive constructs) and visual perception (Gestalt theory) should be understood and considered in conjunction. In Chapter 1, I present underlying descriptive theories to explain how we see and process information before I give recommendations on how to visualise information.

All my research to date is built on the premise that knowing how we perceive and interpret visual information is equally important as understanding how we process and memorise information. Building on this same premise, in Chapter 1 I start by discussing cognition (thinking) theories and principles to show how the combination of graphics with text reduces the mental effort we have to apply to processing and memorising information. By reducing this effort, we can then focus more on the content and the task they have to complete, instead of trying to decode the way information is presented to us.

I then move on to explaining, discussing and illustrating principles and tools in visual perception (seeing) that designers can use to facilitate the simplification of visual complexity. For us to understand the world as it is presented to us, which is visually compelling but complex, we need to find mechanisms to simplify it. In this case, the simplification of information. Knowing the various effects of these principles and tools on perception, and interpreting them adequately, is of great benefit to designers.

In sum, Chapter 1 gives a digested account of **underlying theories for information visualisation** that encompass cognition and visual perception to equip designers with the knowledge needed to better understand how humans perceive, process and retain information.

CHAPTER 2 | FROM THEORY, TO RESEARCH, TO PRACTICE

While some guidance is available for the design of information visualisation, there is still a great need for guidelines that bring research and practice together and that take visual perception and cognition into account. Such guidelines are both necessary and useful, and as with any guidelines, they are open to interpretation and can be adjusted to the specific information context being dealt with.

From Theory... | In Chapter 2, at the start of each design feature (typography, colour, graphics, layout, data visualisation), I explain why considering such features within information visualisation is important and why there is a need to follow certain guidelines. **To Research... |** This is complemented and strengthened by research findings that give evidence of best practice.

I then present guidelines that emerged from a thorough and in-depth investigation. These are a combination of guidelines from research-based and practice-based guidelines. I extracted research-based guidelines from academic papers involving experimental testing, academic papers with a more theoretical focus, academic papers reviewing and discussing existing research, etc. When not presenting an experimental study, such academic papers also needed to have a good set of academic references to support their analysis and discussion to count as research. As for practice-based guidelines, I extracted them from several books, online articles, blogs, etc. These are guidelines that give recommendations based on experience from practice that can complement the guidelines from research, or replace them when they do not exist.

However, providing mere guidance on how to visualise information is not enough to achieve positive and high impact. Even cases where well-researched guidelines are presented can be misinterpreted and do more harm than good, as discussed above. Scott et al.'s (2017) paper 'How to make an engaging infographic?' is an example where theory, research-based principles and Gestalt principles were gathered by health experts and followed, only to result in a very poorly designed infographic. The infographic actually goes against some of the principles listed in the paper. For example, the paper lists 'restrict colour', but colour is used excessively in the infographic and in combinations that hinder legibility; it lists 'align elements', but most elements are misaligned – including the sentence 'align elements'.

To practice... | To avoid this same problem and to maximise the application of the guidance provided in this book, and its success, in addition to the above, I present top guidelines with visual aids that always show the dos and the don'ts. At the end of each section of Chapter 2, I provide a practical example of the application of these guidelines to an entire infographic. Again, two versions are presented: a version showing and explaining the don'ts, and another version of the same infographic showing and explaining the dos.

CHAPTER 3 | FROM STATIC, TO INTERACTIVE, TO MOTION

The benefits of both interactive and motion graphics have been identified in the literature, and these two forms of communication are becoming more and more popular. However, there are very few guidelines and empirical foundations about how to best design them. Moreover, the few studies available rarely refer to issues of cognitive load and how to reduce cognitive overload. With this in mind, I have dedicated an entire chapter (Chapter 3) to interactive infographics and motion graphics that will focus on filling this gap in knowledge in the field of Information Design. I will also explain in more detail the features, types and various components that are part of interactive infographics and motion graphics and that can be manipulated to design the optimum output. Finally, I propose and discuss a new strand of information visualisation, for which I have coined the term 'intermotion graphics', which brings interactive and motion graphics together to form animated graphic interactions. For this new strand, I also present a list of principles that, due to the lack of research in this area, were informed by a handful of practice-based sources and by my work as an information design researcher. Such principles should be considered when aiming for the optimum intermotion graphics design.

CHAPTER 4 | USER-CENTRED RESEARCH METHODS FOR VISUALISATION

Chapter 4 presents, explains and discusses user-centred research methods for information visualisation that are then put into practice in Chapter 5. This will equip readers with relevant practical research skills and tools to use when developing information visualisation. Eleven different methods are presented following the stages that I define as being key for a user-centred research and design process: 1. Discovery; 2. Exploration; 3. Development; and 4. Evaluation. Each method is explained in detail, and the main points that should be considered when using such methods are discussed. A step-by-step guide of how to conduct each method is also given in the form of an infographic. At its core, Chapter 4 illustrates the importance of designing for the user and with the user, and how this can only be achieved if research informs the creative design process at every stage. Moreover, this chapter also shows how research can help us produce accessible, clear, easy-to-use and engaging information outputs that withstand exposure in real-life scenarios.

CHAPTER 5 | FROM THEORY, TO RESEARCH, TO PRACTICE... AND BACK

Chapter 5 is the culmination of everything I have discussed in the book by implementing all that was learnt and then giving back. This means providing a series of research and design studies, conducted by me and my research and design teams, that give **back** solid research-based guidance. This also means, yet again, filling in a gap in knowledge. As I discussed earlier in this section, despite the great benefits of information design and visualisation, research studies in this area are scarce and only a few academic papers are available. Moreover, literature on information visualisation is largely unscientific and limited to a few books providing practice-based recommendations that derive from tacit knowledge acquired through practical experience. Design practice can certainly help us understand how design guidelines can be applied to produce effective design solutions. However, without empirical evidence and user-centred approaches, we cannot validate our design solutions and ensure that they fulfil their primary objective to communicate with the end user in a clear, accessible and engaging way.

By presenting six studies on the effectiveness of information visualisation to communicate complex and/or large amounts of information, I am providing empirical foundations for the design of information visualisation with clear application to practice.

1

Underlying Theories for
Information Visualisation

INTRODUCTION

Humans have a limited attention span. Research shows that information visualisation grabs users' attention and increases engagement. Moreover, it helps minimise inattentional blindness and users ignoring important information. Humans also have limited memory capacity. Information processing and assimilation of information take place in the working memory, which is extremely limited. Research further shows that visualisation of information can have a positive effect on users in relation to cognitive matters.

Information visualisation combines knowledge from various disciplines, with graphic (visual) design and cognitive science being the most relevant to the content covered and discussed in this book. Authors such as Mol (2011) have explained how information visualisation takes advantage of the pathway between the human eye (visual perception) and the mind (cognition) to facilitate users to see, explore and understand large amounts of complex information. In addition to the cognitive domain, the affective domain also plays an important part in this process (i.e., to like or dislike something), with cognitive science showing that we learn better and understand more if information is well designed and is aesthetically pleasing to the eye.

Compared to text-dense information, infographics, for example, provide information through a combination of text, symbols, colours, graphics, and other elements, which is more attractive but also easier to digest and assimilate. In most instances, text is still required as an integral and dominant element of visualisation. Moreover, according to the dual coding theory of Allan Paivio (1986), our memory has two codes (i.e., two channels): one that processes verbal stimuli (words) and one that processes visual stimuli (images). While these two stimuli are stored independently, they form links between each other (i.e., between text and images), which in turn facilitates and enhances the processing and retrieval of information. This rationale further strengthens that information visualisation is the best design approach precisely because it combines both visual and textual elements, making it more accessible than other forms of communication across a wider range of users.

As in any field of research, there are different schools of thought on what is the best form of visualisation of information. For example, some researchers argue that decorative elements (what is defined as 'chart junk') are unnecessary and only serve to distract users from the message. Other researchers argue that attractive information engages and draws users in (which is in line with the cognitive affective domain).

Regardless of the different schools of thought, all in all, for information visualisation to effectively enhance communication, it should be clear, accurate, relevant, actionable and visually pleasing. This can be achieved when principles of perception and cognition are also taken into account. Such an approach creates a balance and harmony between the various elements of an information visualisation (text, colour, graphics, layout).

In this chapter, I discuss and explain in a digested way underlying theories that are of great relevance to help us understand human cognition and perception, which will in turn inform our design decisions during the development of information visualisation.

1.1 | COGNITION
(THINKING)

Memory

Cognitive Load

Cognitive Load Effects

15 Cognitive Principles in Visualisation

MEMORY

There are three types of memory that process visual information in our brain: iconic memory, working memory and long-term memory. Short-term memory and external memory relate to working memory, as explained in this section.

ICONIC MEMORY – Information usually remains in the iconic memory for less than a second before being forwarded to the working memory. Thus, at this first stage of memory, information is processed extremely quickly, and this process is nothing more than an automatic and unconscious perceptual processing – pre-attentive visual processing. Here, several features are detected, making these the elements that stand out when we first look at information, such as:

- Colour (hue and intensity);
- Form (length, width, orientation, shape, size);
- Location of elements in a 2D space.

Consequently, if something important needs to stand out in an infographic or chart, the information should be encoded using a pre-attentive attribute that has good contrast with the surrounding information. The same is true if a particular set of elements needs to be seen as a group, where a pre-attentive attribute can be assigned to such elements.

WORKING MEMORY – Information is instantly moved to the working memory, which is temporary and has limited storage capacity. At this second stage of memory, what our brains recognise as useful will be combined into meaningful chunks of information. According to the chunking principle, the cognitive load can be reduced if visualisations are presented in chunks. (Cognitive load is explained in more detail in the next section.) However, only three to five chunks of information can be stored at one time in our working memory (previous studies have identified up to seven and even 11 chunks but this was a rough estimate; recent studies have shown three to five chunks as being a more precise capacity limit). Nonetheless, these can contain a good amount of information. Therefore, for new information to be included in our working memory, something that is already there will be either forgotten or forwarded to our long-term memory. In a learning context, for example, learning is affected when the working memory capacity is exceeded, and cognitive processing becomes ineffective. In the context of data visualisation, if a legend for a chart contains, for example, a different colour or icon for eight, 10 or 12 different datasets, this will put a lot of strain on the user. Because users' working memory cannot take in more than three to five chunks of information, users will have to go back and forth between the legend and the chart to process all the different pieces of information. When it comes to the size/quantity, users are able to hold more information in their working memory if a large amount of information is coherently and consistently chunked. An example is the fact that charts are able to communicate a large amount of information because it is perceived all at once and as a meaningful pattern, although it actually consists of hundreds and thousands of values. However, tables are only used to look information up; it would be impossible to take a series of numbers and chunk them together meaningfully to store in our working memory.

For clarity, it is important to note that working memory and short-term memory are sometimes used interchangeably. After in-depth research to reduce the confusion between short-term memory and working memory, Cowan (2008, p.13) concludes that 'the distinction between short-term memory and working memory is one that depends on the definition that one accepts' and 'may be [also] a matter of semantics'. In pragmatic terms, as defined by Budiu (2018), working memory is task oriented, while short-term memory represents the brain process that permits the storage of information.

EXTERNAL MEMORY – To make sure the user experience is pleasant, designers must limit the demand put on the users' working memory, i.e., facilitate users' access to the information they need. However, even with a good design, some tasks are naturally more complex than others. Budiu (2018) suggests external memory as a means to help with the limitations of working memory. Budiu compares external memory with the solution of providing pen and paper, so that users can write down the information needed to complete a task without having to store it in the working memory. As Budiu puts it, 'the paper acts as a physical scratchpad, a "fake" working memory' (Budiu, 2018). Types of external memory include any tools or UI (User Interface) features that permit users to save and/or access information needed during a task. Examples listed by Budiu (2018) as forms of external memory include:

- Comparison tables, when shopping online to select products of interest to compare features and pros and cons;

- Shopping basket online, when saving products of interest to then decide at the end which one to buy;
- Page parking, when saving interesting items for a future inspection without interrupting the task of selection.

LONG-TERM MEMORY – When we finally decide to store information for later use, we send chunks of information to our long-term memory. Long-term memory is extremely important to visual perception because it is where our ability to recognise visuals is held. Therefore, in terms of visualising information, iconic memory and working memory is where we want to make sure that our information is perceived and processed adequately.

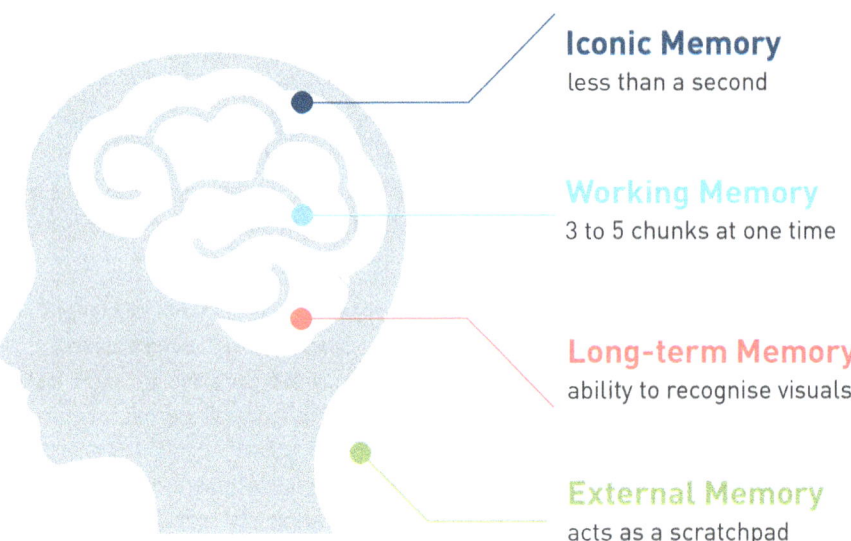

Iconic Memory
less than a second

Working Memory
3 to 5 chunks at one time

Long-term Memory
ability to recognise visuals

External Memory
acts as a scratchpad

Figure 1   Types of memory that process visual information in our brain

COGNITIVE LOAD

When learning new information, we manipulate that information in the working memory before sending it to and storing it in the long-term memory. During this stage of processing, we will be subject to different types of cognitive load that will limit the amount of information we can process during a period of time. (Cognitive load is defined in this book as the effort needed in the working memory to process information.) This means that if we have too much information to process or the information is difficult to understand, our limited working capacity can be overloaded – cognitive overload – and impair comprehension.

INTRINSIC COGNITIVE LOAD – Refers to the information in the working memory relevant to the task and to factors intrinsic to the information content (such as the conceptual difficulty of the information). It occurs when the information presented is complex and various elements must be processed at the same time. It is also dependent upon the level of interactivity between the nature and complexity of the material and the level of expertise of the learner. Therefore, to decrease the intrinsic load, users must be given information with a low level of interactivity (low in complexity and within users' expert knowledge), or one to two concepts that are independent and can be assimilated and memorised.

EXTRINSIC COGNITIVE LOAD – Also known as ineffective cognitive load and commonly defined as load which detracts from the process of learning. It refers to the information in the working memory irrelevant to the task and to factors external to the conceptual message but related to the way the message is designed (font, graphics, colour, layout, etc.). In addition, extrinsic load occurs when there is too much information and the users have to split their working memory attention between several sources of visual information. This discourages attention, but attention is crucial for the brain to integrate the information together to gain understanding and build knowledge.

Element interactivity also leads to extrinsic load. This refers to the excess use of visual elements such as colour, lines, icons/symbols, shadows, 3D charts, etc., that are external to the content. Any extraneous visual element, such as shadow boxes, can interfere with processing the intended content. Interactivity when not intrinsic to the content can impose an even higher cognitive load.

GERMANE COGNITIVE LOAD – Also known as effective cognitive load, refers to processes that users focus on that are directly relevant to learning. In other words, users need to be active participants and of course mental effort is necessary to understand the information. Naturally, all these three processes increase cognitive load, but germane cognitive load contributes to effective learning and processing of information, instead of interfering with it.

To reduce the cognitive demands imposed on working memory by these three sources of cognitive load, the levels of each source should be adjusted. To this end, designers should focus on reducing both intrinsic and extrinsic cognitive load, while encouraging and facilitating germane cognitive load. At all times, this adjustment should make sure that the overall requirement of working memory is within the capacity of users.

Intrinsic load can only be managed by changing the nature of what is learned or by the act of learning itself. To reduce intrinsic cognitive load, designers can, for example, split information from a complex task into smaller, simpler steps for users to complete the task step by step as opposed to learning the entire task all at once. Or designers can visualise complex text using icons and illustrations that can improve comprehension and recall by allowing concepts to be more explicitly depicted than if only text is used.

Designers should be more aware of extrinsic cognitive load, while also recognising that it is the one that is more easily manipulated. The organisation of text into simple list formats, or limiting the number of colours, fonts, shapes and lines, can reduce extrinsic cognitive load and in turn increase understanding and recall.

Information needs to have plenty of germane cognitive load. Designers can use mnemonics to promote germane load. Imagery and visualisation are excellent mnemonics because we remember images much more easily than words, and designers can therefore visualise some text to stimulate germane load. Chunking information is another way of promoting germane load. This mnemonic device is very important in the creation of infographics. As previously discussed, our brains can only process certain amount of information at a time. By breaking down larger pieces of information into small, easy-to-remember chunks, we are simplifying information that otherwise can be too complex to understand and recall.

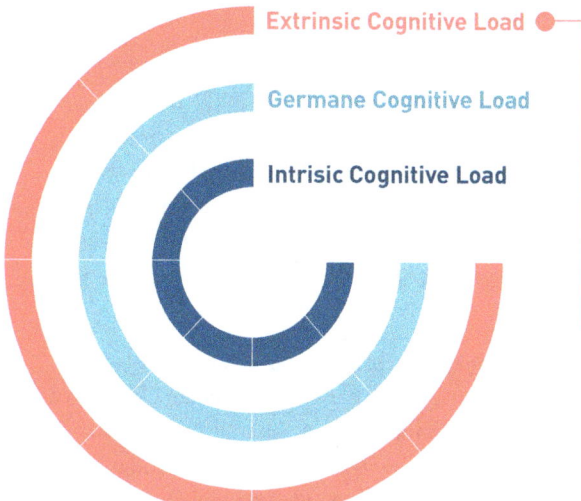

Extrinsic Cognitive Load ●

Germane Cognitive Load

Intrisic Cognitive Load

Related to how information is designed. The one designers should be aware of and manipulate. To reduce it:

- organise text
- limit colours
- limit fonts
- limit shapes
- limit lines
- no shadows

Figure 2   Types of cognitive load that can limit the amount of information people can process during a period of time

COGNITIVE LOAD EFFECTS

Cognitive load effects should also be considered when designing information. Sweller et al. (1998) have identified several cognitive load effects, but for the purpose of this book this section only discusses those that have a greater effect on information design. Furthermore, to have a more comprehensive and holistic approach with high applicability to design work, this section covers both static and motion visuals.

SPLIT ATTENTION EFFECT – Split attention occurs when users need to gather information from two different sources or places. For example, when placed at a distance from each other or in a non-sequential manner when in the same area, requiring users to find and integrate this information that cannot be understood in isolation. The solution is to replace multiple pieces of information, such as separate pictures and text, with a single and well-integrated piece of information.

An example of split attention in static visuals is when we have text on one page of a textbook and the image corresponding to that text is on another page. Even placing text on the same page but far from the corresponding visual leads to split attention. For example, it is better to have the percentages and identification names in the pie chart next to the corresponding slice, than making users go back and forth between the pie chart and the legend. An example of split attention in animated visuals is when animations that complement each other run simultaneously.

MODALITY EFFECT – This effect is similar to the split attention effect. The difference is that split attention focuses on making the integration of information simpler, while the modality effect focuses on improving the processing ability of working memory. The modality effect takes place when information can be processed through two channels (e.g., visual and auditory) instead of one channel, which in turn reduces extrinsic cognitive load. For an example of static visuals, infographics are the best application of the modality effect principle to combine text and images. This is exactly what infographics do, i.e., they combine text and visualisation in one single graphic design to explain or tell a story. In animated visuals, information can be processed more easily using both auditory and visual modalities, i.e., using spoken explanatory text and making sure that audio narrations appear at the exact same time as the corresponding animation.

REDUNDANCY EFFECT – The redundancy effect happens when multiple sources of information are presented together but would benefit from being processed individually. Instead of requiring users to integrate information, the redundancy effect requires users to determine whether information presented in multiple forms (e.g., communicated through text and sound) is useful to complete the task or is different in content. To reduce extrinsic load through redundancy in static visuals, the use of information with an identical function should be presented only once. For example, we should not read presentation slides verbatim. How frustrating it is when someone asks us to look at a piece of news, for example, and then starts reading it for us at the same time as we are trying to read it ourselves and at a different pace. In animated visuals, there is no need either to narrate information in the exact same way as the text is presented in motion graphics. The motion graphics should present summaries or key works that link what is being narrated to what is being seen on screen. Another example is the use of captions to meet user needs, such as auditory impairment or translation of a foreign language. To remove redundancy, captions should be optional to the users.

15 COGNITIVE PRINCIPLES IN VISUALISATION

Having explained and digested human cognition for the purpose of information visualisation, what does it all mean and how should we apply it to practice and real-life situations? This is a gap in knowledge in the field of information design that needs to be addressed.

To this end, a series of cognitive principles have been put together based on various observations in the literature. These have been defined to inform and support the visualision of information when the aim is to reduce cognitive load. The principles are presented in alphabetical order to facilitate searching when referring back to them. They are also presented at the end of this section in a more digestable form (i.e., in a table) for quick reference.

1 CHUNKING – Elements should be grouped together in a meaningful way. Users will process meaningful units as one chunk of information, as opposed to separate bits of information. This will then help them process and remember the information better. For example, creating subtopics in an infographic is particularly helpful, as it will allow grouping information in chunks.

Figure 3   Cognitive principle – Chunking

2 CONSISTENCY – Constant and common information across various infographics, or pages that are part of the same series of information, should be put in the same relative position. If users know where constant information will be found, locating it will be a much easier and quicker process. For example, placing the heading, reference icons, etc. always on the same area of the page.

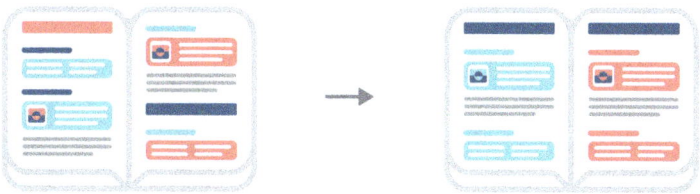

Figure 4   Cognitive principle – Consistency

3 CONTINUED ENGAGEMENT – When focused learning is the goal, it is important to grab users' attention and keep them engaged for longer. Longer engagement with the information can be stimulated by using elements that can be interpreted more quickly than text and are also attractive to the human eye. For example, using illustrations and icons. The best way to achieve this relationship between text and images is by placing the images to the left of the text because the eye will be drawn to the image before the text – just as the natural flow of reading is from left to right. This reduces effort in having to backtrack information.

Figure 5   Cognitive principle – Continued engagement

4 CONTINUITY AND PROXIMITY – By keeping elements together that relate to each other and in sequence, we avoid split attention and the unnecessary effort users will have to make to find and integrate the information. The most typical example is to make sure that images that refer to a specific part of the text are close to it and not on the following page or too far down the page.

Figure 6   Cognitive principle – Continuity and proximity

5 ELEMENT INTERACTIVITY – Any elements that are not intrinsic to the content only increase mental effort. While the overuse of visual elements should be avoided, when multiple elements of information do need to interact, they should be designed in such a way that users are able to assimilate them simultaneously. For example, use the same colour for an icon and correspondent text.

Figure 7   Cognitive principle – Element interactivity

6 EMPHASIS AND SALIENCE – Emphasis can be used to facilitate finding information, such as using different colours or sizes of typeface (but all in moderation). This is quite important for iconic memory, i.e., to grab users' attention at a glance.

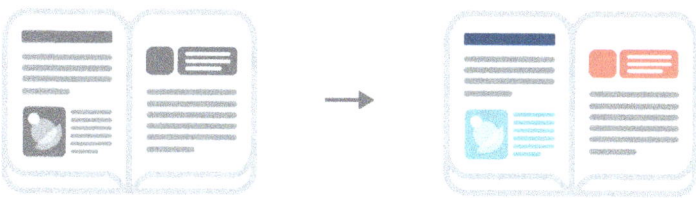

Figure 8   Cognitive principle – Emphasis and salience

7 FAMILIARITY AND CONNECTION – The way information is visualised should build on the knowledge that users might have regarding icons, colour associations, etc. It is therefore better to use familiar elements. But if an icon or colour coding are likely to be unfamiliar and/or are being learnt for the first time, then a label can be added to help the user understand it.

Figure 9   Cognitive principle – Familiarity and connection

8 FOCUS – When distracting information cannot be avoided, the strategy should be to use visualisations that help guide and focus user attention, in turn reducing distractions. This can be achieved by using, for example, clear labels that quickly clarify the content of the information, or visual cues such as arrows pointing directly to the information that users should focus on.

Figure 10   Cognitive principle – Focus

9 HIERARCHY AND ORGANISATION – Hierarchy is used to group and present information in sequence so that users can easily understand its importance, relevance, order, priority. A hierarchy should therefore be presented in the order users are likely to use it or that we would like them to use it. That is, if we organise information in a way that is compatible with how users store or should store information, then it will be more easily processed, retained and recalled. This is the particular case of information that instructs users to take action.

Figure 11   Cognitive principle – Hierarchy and organisation

10 NUMBER OF VISUALS – Although visualisation makes information more accessible and quicker to interpret, using too many visuals is counterproductive and increases mental effort as users are required to discern which elements are relevant. Any unnecessary elements that are not important for the processing and understanding of the information should be removed (e.g., logos, decorative elements, unnecessary grid lines).

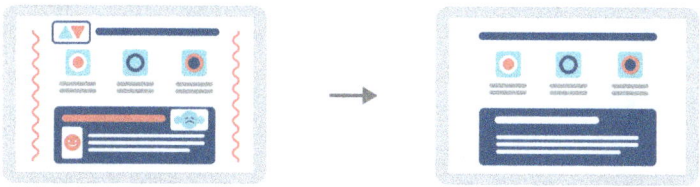

Figure 12   Cognitive principle – Number of visuals

11 ORDER – Information should be ordered so that the most relevant information comes first, as this will affect how users perceive the subsequent information. Moreover, what is presented first usually sets an expectation. This is the case of the relationship between a title in an infographic and the subsequent headings, or an introductory section, followed by the main content and final summary.

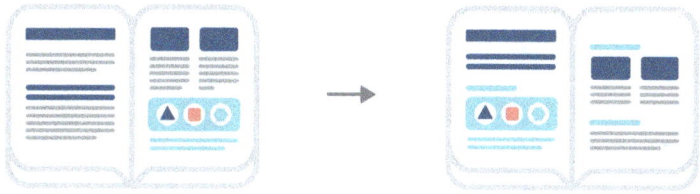

Figure 13   Cognitive principle – Order

12 RANGE OF CHOICES – The number of choices given to users should be limited. Although it might seem a good idea to give various choices, users will struggle to make a decision when faced with too many options. This will not only increase mental effort as it will waste users' time, it also goes against the whole premise in information visualisation to simplify and reduce effort.

Figure 14   Cognitive principle – Range of choices

13 REASONING – Reasoning refers to the action of thinking about something in a logical way. The use of colour-concept associations and colour coding are good ways of helping the user to quickly identify categories within information. For example, the traffic light system used in food labelling, terror threat levels system, international travelling in times of pandemic, etc.

Figure 15   Cognitive principle – Reasoning

14 SIMPLICITY – Once users locate the information, for it to be easily understood it should be as simple as possible. An example of simplicity in information visualisation is the use of icons/symbols that quickly convey the meaning of the information. This is especially important when we want users to understand information meant to alert: danger, emergency, hazards, etc.

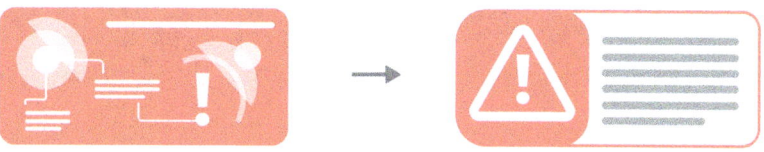

Figure 16   Cognitive principle – Simplicity

15 VISUAL CUES – Visual cues should be used to direct users' attention to specific information, as well as to remind users about the information they are seeing when they need to make sense of more than one piece of information at the same time. For example, when users must remember the type of data they are seeing while also finding patterns, labels and other supporting text elements are very helpful.

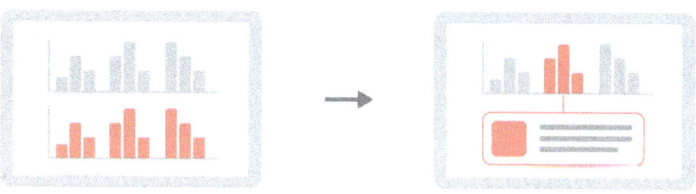

Figure 17   Cognitive principle – Visual cues

In summary, well-designed information should:

1. Engage and promote high-level cognitive functioning, i.e., to gain insight, reasoning and understanding.
2. Attract the users to relevant information (instead of ignoring it).
3. Promote chunking that provides strong retrieval cues that will then be passed onto long-term memory and support reasoning, thinking and decision making.

15 COGNITIVE PRINCIPLES

How to reduce cognitive load in information visualisation?

	Principle	Guidance	Example
1	Chunking	Group elements together in a meaningful way.	Create subtopics and group each topic as a chunk of information.
2	Consistency	Common information should be constant in a series of infographics.	Place headings, reference icons, etc. in the same place.
3	Continued engagement	Longer engagement is needed for focused learning.	Use elements that are interpreted more quickly: illustrations, icons.
4	Continuity and proximity	Elements that relate to each other should be together and in sequence.	Place images that relate to a specific part of the text, close to it.
5	Element interactivity	The user should be able to assimilate multiple items simultaneously.	Use the same colour for an icon and correspondent text.
6	Emphasis and salience	Emphasis should be used to facilitate information searching.	Use different colours or sizes of typeface, but all in moderation.
7	Familiarity and connection	Information visualisation should build on the knowledge of the user.	Use familiar icons and colour associations; or a label if unfamiliar.
8	Focus	Distractions should be reduced using visualisations to guide focus.	Use clear labels or visual cues such as pointing arrows.
9	Hierarchy and organisation	Should be used to easily understand importance, relevance, order, priority.	Present information in the order users are likely to/should use it.

	Principle	Guidance	Example
10	**Number of visuals**	With too many visuals the user has to discern which elements are relevant.	Remove irrelevant visuals: logos, decorative elements, grid lines, etc.
11	**Order**	What is presented first affects how subsequent information is perceived.	Write the infographic title to set the expectation of the main content.
12	**Range of choices**	Users struggle to make a decision when faced with too many choices.	Limit the number of choices given to the user.
13	**Reasoning**	Reasoning is the action of thinking about something in a logical way.	Use colour associations and colour coding to help find categories.
14	**Simplicity**	Once located, information should be as simple as possible to understand.	Use icons/symbols that quickly convey danger, hazards, emergency.
15	**Visual cues**	Should be used to direct users' attention to specific information.	Use labels for when remembering data while also finding patterns.

Figure 18 15 Cognitive principles

1.2 | **VISUAL PERCEPTION**
[SEEING]

---

10 Gestalt principles in visualisation

- Simplicity
- Proximity
- Similarity
- Enclosure
- Closure

- Continuity
- Connection
- Figure-ground
- Focal point
- Common fate

10 GESTALT PRINCIPLES IN VISUALISATION

Ways of reducing cognitive load through chunking include the use of Gestalt principles of grouping. Understanding how we perceive patterns, forms and organisation of what we see is a core foundation for successful information visualisation. Gestalt principles are laws of human perception that show us how to group graphical elements in particular ways to help users interpret those various elements and build relationships between them.

As for the word 'Gestalt', it means an organised whole that is perceived as more than the sum of its parts. Despite presenting Gestalt principles individually in this section, they are not independent of each other. They work simultaneously and it is the application of these principles in combination that has the power to group and enhance the communication of information.

1 SIMPLICITY

We better perceive and interpret abstract and complex information when it is in the simplest form possible. For example, the chart below on the left requires less effort, and therefore is easier to interpret than the chart on the right, because our brains favour things that are clear, simple and ordered.

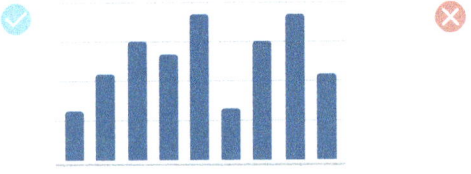 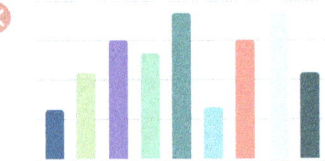

Figure 19   Gestalt principle – Simplicity

2 PROXIMITY

We perceive elements that are close together as belonging to a group. The manipulation of white space is a powerful tool when it comes to organising information and directing the viewers to particular information. For example, in bar charts, bars that are clustered will be seen as a group.

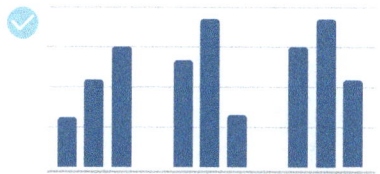 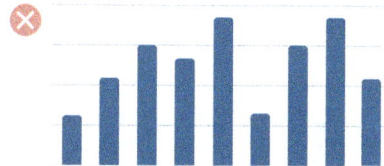

Figure 20   Gestalt principle – Proximity

3 SIMILARITY

We perceive elements that have the same typeface or are similar in size, shape, colour (both hue and intensity) or orientation as being part of the same group. This principle works very effectively as long as the differences are only a few and clearly distinct from one another. For example, in a bar chart, if the bars are the same colour it will indicate to users that the bar values should be compared (left bar chart below). If different colours are used, it will impose extra cognitive load and may indicate that each bar belongs to a unique attribute, and therefore do not need to be compared (right bar chart below).

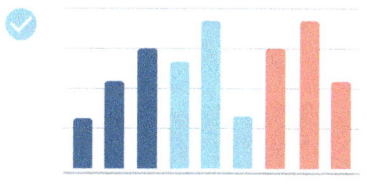 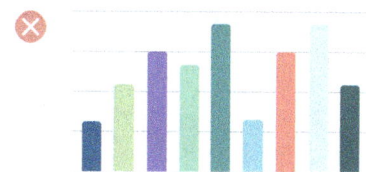

Figure 21   Gestalt principle – Similarity

4 ENCLOSURE

We perceive elements as belonging to the same group when they are enclosed in a way that seems to create a boundary around them (e.g., border, common field of colour or shade, as illustrated in the chart below). Enclosure is the strongest approach of visual perception to group elements.

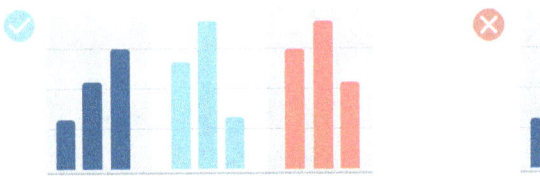

Figure 22   Gestalt principle – Enclosure

5 CLOSURE

We perceive open structures as closed, complete and regular if there is a way that allows us to reasonably interpret them as such. For example, in a chart, we do not need complete borders to define a space. This is particularly relevant for the x- and y-axes, where it is actually preferable to define the area of the chart by using one single thin line (y-axis line on the left and x-axis line at the bottom), rather than with heavy lines around the entire area that will only create a boxed chart (as illustrated below).

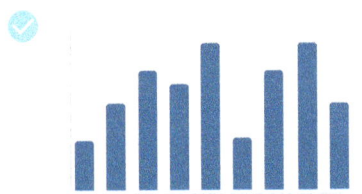 

Figure 23   Gestalt principle – Closure

6 CONTINUITY

We perceive elements as part of a whole if they are aligned or seem to form a continuation of one another. For example, in a horizontal chart, aligning all the bars to the left, even without a line to indicate the y-axis, makes it obvious that they share the same baseline and are part of the same chart. In stacked bar charts this is not the case for the subsequent bars (shown below).

Figure 24   Gestalt principle – Continuity

7 CONNECTION

We perceive elements that are connected (e.g., by a line or point liner) as belonging to the same group. Connection is a stronger tactic than proximity or similarity (colour, size and shape) because it explicitly shows the link between elements. For example, using a line to connect points in a chart is an effective approach (as shown below). It is very difficult for our eyes to connect points without lines.

Figure 25   Gestalt principle – Connection

8 FIGURE-GROUND

We perceive elements as either figure (the element in focus) or ground (the background on which the element rests). When dealing with information, and particularly in situations of time pressure where scanning might be used, it is crucial to be able to determine at a glance what is important (figure) and what is secondary (ground). Therefore, good contrast between the foreground and background should be ensured so that charts are more legible. Looking at the charts below, the chart on the right has additional cognitive load due to the lower contrast between the bars and background. This means that users will take longer to determine which elements are figures, i.e., that communicate data and need immediate attention, and which elements are ground, i.e., not as important and can be left to be interpreted later.

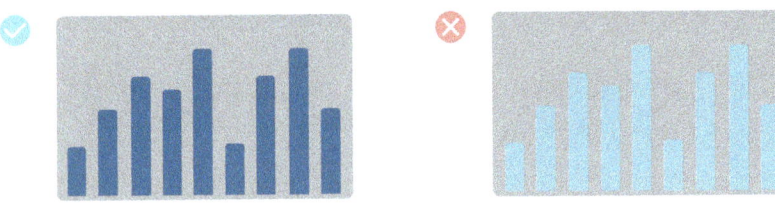

Figure 26   Gestalt principle – Figure-ground

9 FOCAL POINT

Our attention will be grabbed by a point of interest, emphasis or difference. Therefore, distinct elements will be perceived as creating a focal point, and distinct features, such as colour, size and shape, can be used to highlight and create focal points. For example, in a bar chart, if only one bar is presented with a different shade or colour, users' attention will be automatically directed to that bar first. This is a good technique to use when an important data value needs to be highlighted among the rest (e.g., a significant finding).

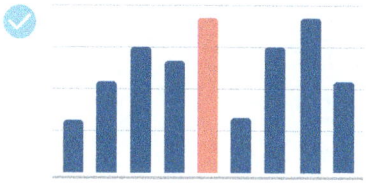
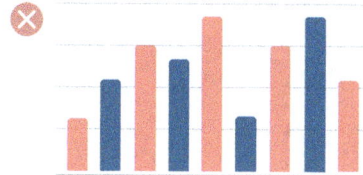

Figure 27   Gestalt principle – Focal point

10 COMMON FATE

Lines that move in the same direction are perceived as belonging to the same group, i.e., they have a common fate. For example, in a line chart, lines trending in the same direction will be easily interpreted as belonging to the same or common dataset. Therefore, lines moving in opposite directions should only be used if the intention is to show exactly that, i.e., opposite trend.

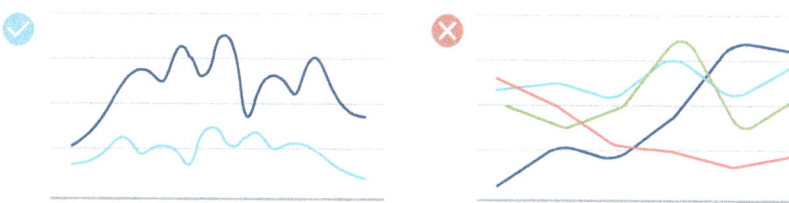

Figure 28   Gestalt principle – Common fate

In summary, it is important to have Gestalt principles in mind when designing information to be able to:

- Understand how information might be perceived;
- Apply the principles as an organised whole (as needed);
- Take advantage of the benefits that these principles bring to information visualisation.

10 GESTALT PRINCIPLES

How to reduce cognitive load through the use of Gestalt principles of grouping?

	Principle	Guidance	Example
1	Simplicity	Complex and abstract information should be in the simplest form.	Avoid information with too many colours, disordered, cluttered, etc.
2	Proximity	Place elements that belong to a group close together.	Use white space to organise and direct the viewers to information.
3	Similarity	Elements in a group should be similar in size, shape, colour, etc.	Use only a few and clearly distinct differences between elements.
4	Enclosure	Enclosure is the strongest approach of perception to group elements.	Enclose elements in a way that creates a boundary around them.
5	Closure	We can make open structures to be perceived as closed.	Instead of complete borders to define a space we can use 2 lines.
6	Continuity	Group by aligning elements or make them a continuation of one another.	Use different colours or sizes of typeface, but all in moderation.
7	Connection	Group elements by connecting them, which is stronger than proximity.	Use lines or point lines to connect elements.
8	Figure-ground	Elements are perceived as figure (imporant) or ground (secondary).	Ensure good contrast between the foreground and background.
9	Focal point	Attention is grabbed by a point of interest, emphasis or difference.	Use distinct features to create a focal point (e.g., colour, size, shape).
10	Common fate	Elements that move in the same direction belong to the same group.	Line chart – lines belong to a data set if trending in the same direction.

Figure 29   10 Gestalt principles

OVERALL CONCLUSION

The overall objective when using visualised information is to reduce cognitive load, i.e., the mental effort that users have to make when processing the information. By reducing cognitive load, we are allowing users to interpret and assess large amounts of information as quickly as possible, with ease and effectively.

Infographics are particularly relevant in this context because the use of graphics together with text reduces the cognitive load. Moreover, an effective infographic allows users to spend more time focusing on the content, instead of trying to decode the way information is displayed. As designers, we should ensure that only cognitive load that is necessary to process information efficiently is imposed on users.

One of the cognitive principles of great importance to information visualisation is chunking. Gestalt principles promote grouping and should therefore be used in connection with cognitive principles, in particular chunking, to enhance comprehension of information and reduce cognitive load.

WANT TO LEARN MORE ABOUT THIS TOPIC?

Cairo, A. 2013. *The Functional Art: An Introduction into Information Graphics and Visualization*. **Berkeley, CA: New Riders.**

This book dedicates Part II (one of four parts of the book) to cognition, starting with the eye and what Cairo refers to as the 'Visual brain', then moving to visual perception, and finishing with memory.

Few, S. 2012. *Show Me the Numbers*. **Burlingame, CA: Analytics Press.**
and
Few, S. 2013. *Information Dashboard Design: Displaying Data for at a Glance Monitoring*. **Burlingame, CA: Analytics Press.**

Both books give a good and clear account of cognition and perception. They bring together Few's papers, published over the years, making an excellent contribution to information visualisation. Despite the focus being on data visualisation, the way the content is discussed makes it directly applicable to any type of information visualisation.

Kosslyn, S.M. 2006. *Graph Design for the Eye and Mind*. **New York: Oxford University Press.**

Although this book focuses on specific visual displays such as graphs, like Few's books, its content is applicable to the design of other visual displays. The entire book is dedicated to how we perceive and comprehend graphs, and by extension, how our eyes and brains process visual information. The book is also well supported by the scientific literature, which makes its content highly reliable.

Lipton, R. 2007. *The Practical Guide to Information Design*. **Hoboken, NJ: John Wiley & Sons.**

This book starts with a chapter that focuses on the audience. It is in this chapter where Lipton discusses visual perception and the various Gestalt principles, only brushing on cognition: memory and recall. In part, Lipton continues to focus on the perception, with text, colour and symbols being the focus. This is therefore a better book to inform us about visual perception.

Mollerup, P. 2015. *Data Design: Visualising Quantities, Locations, Connections*. **New York and London: Bloomsbury.**

What is interesting about this book is that it is short in content but what is covered is very direct, clear and well explained. It also benefits from simple graphics that helps us visualise and understand the content better.

Pettersson, R. 2021. *Cognition*. **Esse, Finland: Institute for Infology.**

I consider Pettersson's work to be the encyclopaedia of information design. Anything we need to find out about information design, Pettersson has discussed it in one of his publications. What I admire most is Pettersson's strategy to create online versions of his work that he updates very regularly. He is on top of all that is published in the field of information design. *Cognition* is one of 12 books; book 8 and covers attention, perception, the brain and processing. The content is easy to understand and well supported by the scientific literature. But I strongly recommend all 12 books, as they are all an outstanding contribution to knowledge in the field of information design.

Ware, C. 2004. *Information Visualization*. **San Francisco, CA: Morgan Kaufman Publishers.**

This is a denser book to read but necessary for those wanting to delve more deeply in the area of information visualisation. Cognition and perception are well explained throughout the book.

RELEVANT REFERENCES

MEMORY

Authors and researchers who have focused on memory and that have informed this section, include: Bettman et al., 1986; Sweller, 1994; Few, 2004a and 2012; Patterson et al., 2014; Lyra et al., 2016; Tetlan and Marschalek, 2016; Coyle et al., 2017; Majooni et al., 2017; Budiu, 2018.

COGNITIVE LOAD

The cognitive load theory has been explained by several authors and researchers and of particular relevance for this section are: Sweller, 1994; Sweller et al., 1998 and 2011; Artino, 2008; Wilson and Wolf, 2009; Tetlan and Marschalek, 2016; Clinton et al., 2017; Emory, 2019.

COGNITIVE LOAD EFFECTS

Authors who have focused on cognitive load effects and informed this section include: Sweller et al., 1998; Artino, 2008; Wilson and Wolf, 2009; Roodenrys et al., 2012; Tetlan and Marschalek, 2016; Emory, 2019.

15 COGNITIVE PRINCIPLES IN VISUALISATION

The 15 cognitive principles were identified reading the work of several authors and researchers. I was particularly inspired by the work of: Bettman et al., 1986; Patterson et al., 2014; Coyle et al., 2017.

10 GESTALT PRINCIPLES IN VISUALISATION

Authors and researchers of particular relevance for the section on Gestalt principles include: Moore and Fitz, 1993; Lipton, 2007; Mol, 2011; Few, 2012; Ali and Peebles, 2013; Brower, 2014; Knaflic, 2015; Korpela, 2016.

2

From Theory,
to Research, to Practice

INTRODUCTION

Guidelines that bring research and practice together and that take visual perception and cognition into account are both necessary and useful when visualising information. Moreover, guidelines are open to interpretation and can be adjusted to the specific information context being dealt with. Guidelines are for responding to context, not for setting it (Berinato, 2016).

I wanted to bring together existing literature and experiences in good information design to create a series of best practice guidelines. I reviewed literature to:

1. Identify principles for the effective use of information visualisation;
2. Identify obstacles to effective communication;
3. Ascertain and examine case studies where information communication was significant to the outcome;
4. Develop rationales for good practice.

I identified a total of 486 information visualisation guidelines in the literature. These are guidelines from research that were identified in academic papers involving experimental testing, academic papers with a more theoretical focus, academic papers reviewing and discussing existing research, etc. Additional practice-based guidelines, i.e., that give recommendations based on experience from practice, were also identified in several books, online articles, blogs, etc. Although practice-based guidelines are less reliable, as they have not been tested with the end user, they are good to fill in the gaps that research has not been able to fill in yet. They are also important to give a more industry-related approach to any design solution.

Instead of presenting all 486 guidelines, in this chapter I present guidelines in a more applicable and engaging way by doing the following. At the start of each section, i.e., for each design feature, a rationale is given to explain why it is important to consider such guidelines, linking where possible to cognition, perception and user behaviour. Findings from research, i.e., relevant findings generated by the experimental studies reviewed, are also given to reinforce and validate the guidelines as well as to make them more user-centred. Top guidelines for the different elements of information visualisation are then presented by combining text and visuals to illustrate what the guideline means in practice. A final section discusses an example of how theory and research findings can be applied to practice, using infographics as an example.

In sum, in this chapter we move from context to best practice. Best practice is then distilled into key guidelines and actual examples are given at the end to further understanding of how these can be applied in practice. This structure is established to avoid situations like the one described in the Introduction of this book where well-researched guidelines were misinterpreted and did more harm than good.

INFOGRAPHICS VS DATA VISUALISATION

Information visualisation includes two categories: information graphics and data visualisation. Infographics is the contraction of 'information graphics'. Infographics are used to communicate specific information to specific users. They are visual representations of information, data and knowledge and are designed with the goal of communicating complex information in a clearer and more accessible manner than text alone, as well as grabbing attention and interest. The use of text, icons, colours and graphics lend infographics the role of telling the story behind the information and data in a more focused, organised, intuitive and engaging way.

Data visualisations are not infographics, but are featured within infographics. Data visualisation is the visualisation of numeric values with charts, tables and graphics by transforming raw and dense data into visual presentations. It includes clear information based on measurable statistical data.

Data visualisation and infographics, however, exist on a continuum. Of the elements used in infographics, data visualisation is one of the strongest, if not the strongest. First, because it summarises hundreds and thousands of numbers into a digested visual form, and second, because it has direct effect on the credibility and persuasiveness of infographics.

Each section in this book uses infographics to show application to practice and the impact of applying or dismissing information visualisation guidelines. If designed well, effective infographics capture complex ideas, behaviours or knowledge in an easily digestible visual format; deliver maximum information in a minimum amount of time and space; and combine visuals and words to increase consumer comprehension and retention.

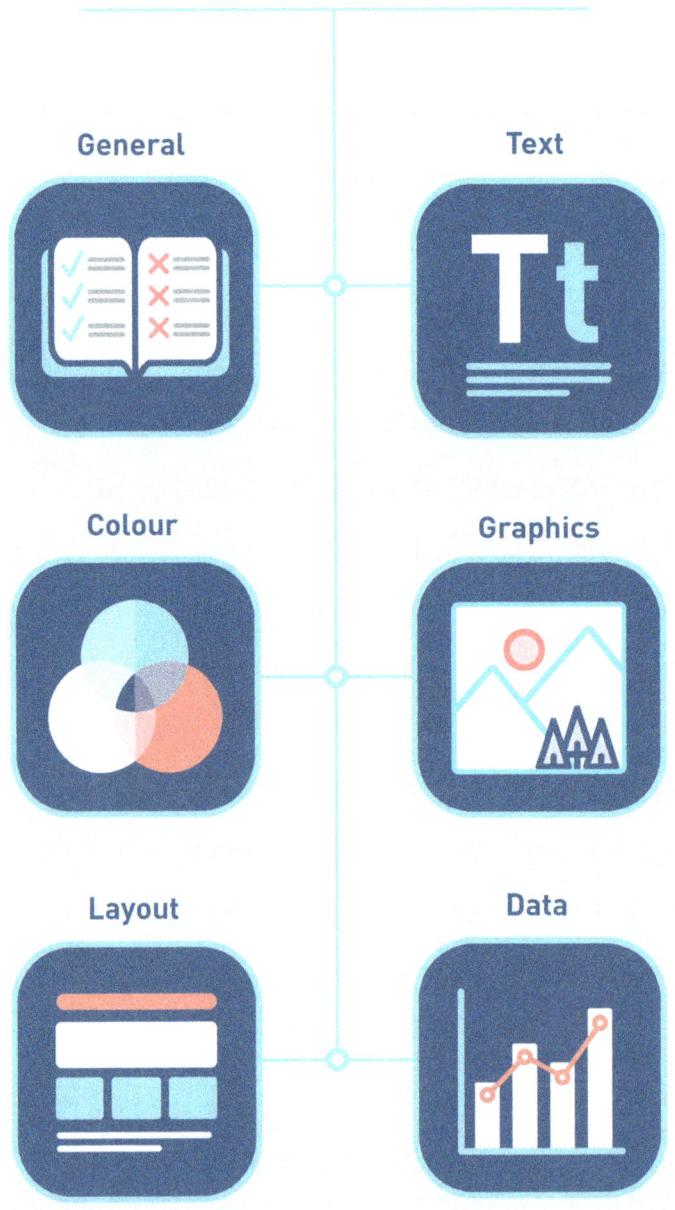

Figure 30 Design guidelines – Areas covered in Chapter 2

2.1 | GENERAL GUIDELINES

WHAT TO CONSIDER BEFORE DESIGNING

Presenting information with infographics can enhance understanding and learning, and influence decision making, i.e., how quickly users can interpret information, but also how the information is interpreted and acted upon. Infographics should therefore be focused, accurate, accessible and clear, instead of overwhelming the users, distracting users from the key message, and leaving users confused and with more questions than the answers the infographic provides.

Infographics also tell a story, even if just a simple data comparison. Graphics, typography, colour and images are the elements used to this end and choosing them inappropriately may result in misunderstanding of the content/story. Moreover, the information needs to be filtered and synthesised, the relationship between the information and elements needs to be established, and patterns need to be represented in a simple way. Good design choices, simple communication of information, and the reduction of elements that distract from the message, require mastery. If these are not achieved, the result will be a loss of information and the production of incomplete and incoherent infographics.

Although appeal is important to grab attention, the main goal of an infographic should be to communicate the information in order to facilitate comprehension and retention. The visual appeal of an infographic alone will not make up for poor design and content.

The way most users interact with infographics is by skimming, hence the need for the key message to be clear and straightforward. Moreover, keeping an infographic to one page ensures that all the elements contribute to the communication of the message at the same time, and that the infographic is easier to scan and then skim.

Therefore, the two essential elements for the development of an infographic are taking into account the target audience and conducting frequent evaluations with users, which should be repeated throughout the design process.

GENERAL BEST PRACTICE FINDINGS

- Infographics were found to facilitate a quick grasp of information, i.e., obtain information in little time and with little effort in situations of time pressure (Zhang, 2017).
- Infographics were considered highly informative, practical, useful and valuable in decision making and business operations. For example, historical data to help with decision making, reports to help with strategy, and tips for better performance (Zhang, 2017).
- The combination of text and graphics in a meaningful and calculated way (as in infographics) was found to be very effective in a variety of learning, instructional and persuasive tasks as well as technical documentation (Zacks et al., 2001; Kendler, 2005; Lyra et al., 2016).
- Infographics within news stories were found to achieve longer viewing times than other images (Holmqvist and Wartenberg, 2005).
- Infographics were found to be intrinsically memorable, with consistency across various groups of people (Borkin et al., 2013).
- The use of visual cues (icons and words) was found to help clarify the meaning of the data and assist in making the right choice (Hibbard et al., 2002; Gerteis et al., 2007; Hildon et al., 2012).
- For comparison of the relative size of two categories, judgement was found to be most accurate along a common scale (simple bar chart), was of intermediate accuracy when assessing length (divided or stacked bar charts), and was the least accurate when assessing angles (pie charts) (Cleveland and McGill, 1984; Simkin and Hastie, 1987; Heer and Bostock, 2010; Schonlau and Peters, 2012).

TOP GENERAL GUIDELINES

1 CLEAR FOCUS AND PURPOSE – Infographics should have a clear focus and purpose, communicate complex information quickly and clearly, and communicate information that is accurate, complete and relevant. Infographics should be efficient, simple (but not simplistic), concise (without leaving important information out), and clear by ensuring that every element in the infographic has a specific purpose and will not be misunderstood by users.

Figure 31   Top general guideline – Clear focus and purpose

2 KEY MESSAGE IN A BLINK OF AN EYE – The key message in an infographic should be communicated in less than five seconds and should be the first information users understand and remember after reading the infographic. When quick comprehension and decision making are at stake, simple and plain infographics should be used and limited to one page (or a maximum of two pages).

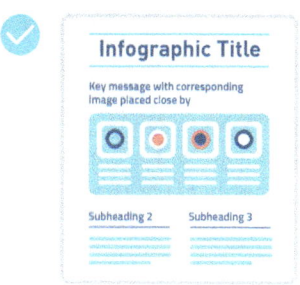 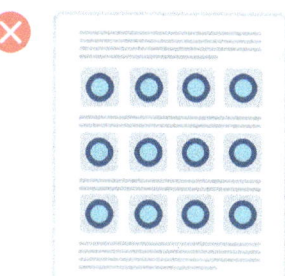

Figure 32   Top general guideline – Key message in the blink of an eye

3 VISUALS FOR ATTENTION AND COMPREHENSION – For attention seeking, infographics can communicate simple messages with the help of visual elements, such as the use of bright colours and relevant images. For wider appeal and to enhance the comprehension and recall of information, embellished infographics can be used with caution and should not be purely ornamental.

Figure 33   Top general guideline – Visuals for attention and comprehension

4 COMPREHENSIBLE TEXT IN AN APPROPRIATE TYPEFACE – Text in infographics should be transparent (i.e., not call attention to itself), easy to read and self-explanatory. In terms of the typefaces used, these should be chosen appropriately for their function, i.e., to convey the infographic message effectively and to fit with the purpose of the text and the infographic. Typefaces with unusual features, typefaces that distract from the text content, and typefaces that have not been tested objectively should be avoided.

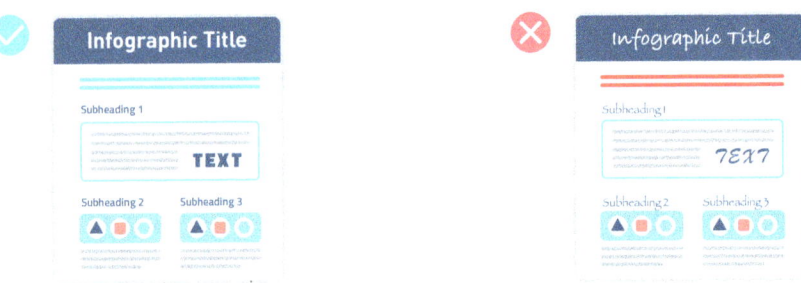

Figure 34   Top general guideline – Comprehensible text in an appropriate typeface

5 DELIBERATE COLOUR SELECTION – Colour selection in infographics should not be subjective or based on personal preferences but informed and deliberate to fulfil the specific needs and purposes of the infographic. Colour can subtly affect mood and opinion, and therefore should be used harmoniously to support effective communication and make users comfortable with the colours used. The meaning of colours also varies in different cultures, regions and contexts. Colours must therefore be taken into account to ensure they do not offend or send the wrong messages to the target audience and have negative consequences.

Figure 35   Top general guideline – Deliberate colour selection

6 EFFECTIVE VISUAL ELEMENTS – Visual elements should be used effectively, be consistent with the function, content and key message of the infographic, and be arranged adequately within the infographic's structure.

 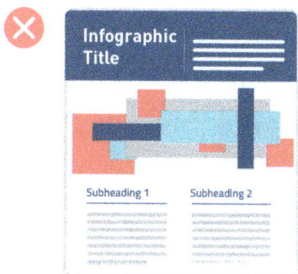

Figure 36   Top general guideline – Effective visual elements

7 APPROPRIATE DATA VISUALISATION – Within an infographic, multiple data visualisation formats should be considered, to suit the type of data to be communicated and the target audience in question. When choosing a visual display for the data, the following questions should be asked: What do users need to find out? What is the best chart for the specific need? Is this chart the easiest for users to interpret?.

 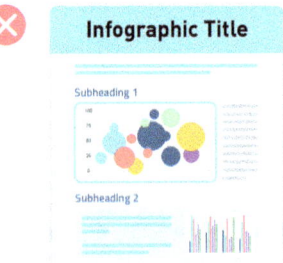

Figure 37   Top general guideline – Appropriate data visualisation

8 USER-FRIENDLY STRUCTURED CHARTS – The content of a chart itself should be organised by:

- Analysing the data thoroughly;
- Classifying it in order of relevance;
- Associating it according to its meaning.

Visual attributes should then be used to help:

- Group information into meaningful sections;
- Establish hierarchy and order of importance;
- Sequence the information according to the order in which it should be read.

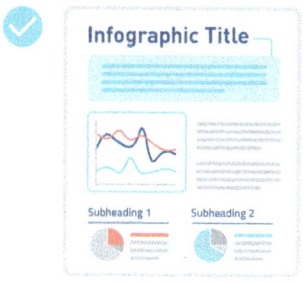 

Figure 38   Top general guideline – User-friendly structured charts

9 ACCURATE DATA PRESENTATION – Data quality, accuracy and simplicity should be ensured by:

- Keeping text, colours, symbols and metrics consistent;
- Carefully selecting a few visual cues to help clarify the meaning of the data;
- Creating clear captions, titles and annotations on how to interpret the visualisation (in particular when less familiar formats are used);
- Ordering the information by rank and relevance;
- Clarifying uncertainty (e.g., clearly labelling associations, comparisons, etc.);
- Displaying data accurately and in context, avoiding distortion and bias.

Figure 39   Top general guideline – Accurate data presentation

10 UNCLUTTERED AND WELL BALANCED – Chart junk and cluttering should be minimised by avoiding distracting patterns, overbearing colours, shading, 3D, unnecessary grids, etc. When considering aesthetics in data visualisation, things to take into account are:

- Be smart with colour – colour should always be used with an intention and used sparingly to highlight relevant parts;
- Pay attention to alignment – elements should be organised in clean vertical and horizontal lines to achieve a sense of unity and cohesion;
- Maximise the use of white space – margins should be preserved by not including unnecessary elements simply because there is space.

Figure 40 Top general guideline – Uncluttered and well balanced

11 TARGETED AT THE NEEDS OF THE USER – The needs and expectations of the audience should be considered at all times and the target audience should be involved in the design, evaluation and dissemination of information. Information visualisations should be validated by usability and performance studies and user feedback. For example, once a few initial infographic designs have been developed, they should be tested by showing them to users and ascertain:

- Where users focus;
- What information users are able to find and how quickly;
- What feedback users give;
- What do users struggle with;
- What questions do users have.

An iterative process of test, refine and test again, should be followed, which will confirm whether the right design is being developed and what changes are needed.

 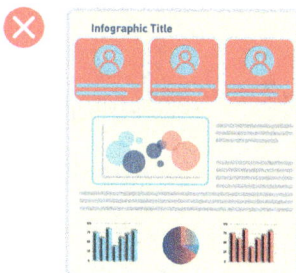

Figure 41   Top general guideline – Targeted at the needs of the user

APPLICATION TO PRACTICE

 DON'T

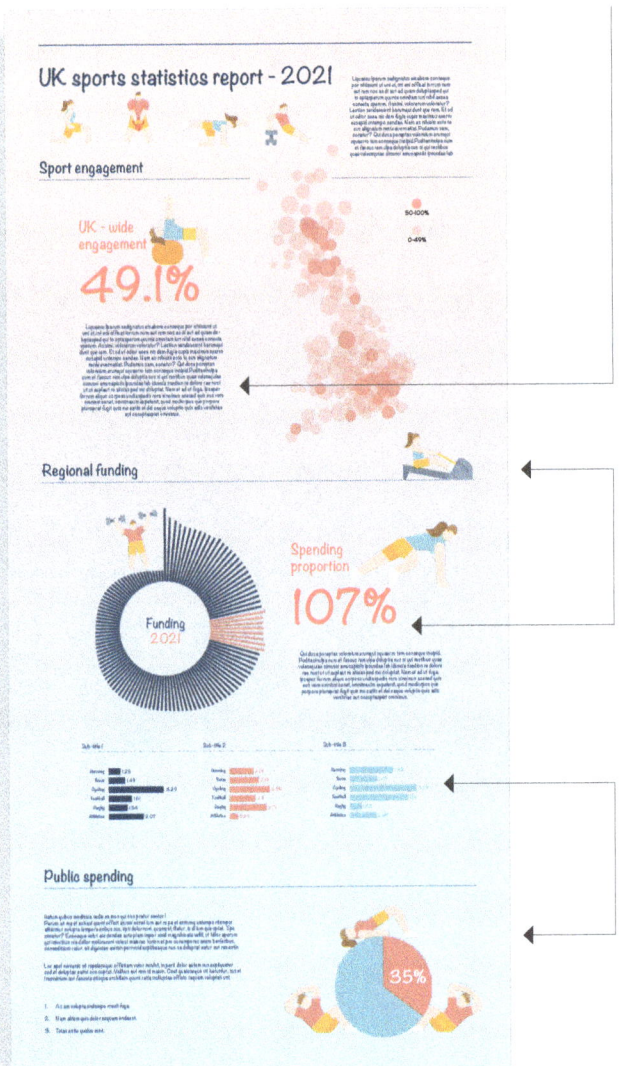

Figure 42 Poor application of general guidelines

Text problems | Overall, text has been poorly designed in this infographic. The font used is too elaborate for the narrative text (better suited for titles and headings). Moreover, narrative text is small and legibility is further impaired because the text is also bold and justified to the centre. These three practices impair legibility and speed of reading. Centred text is particularly problematic because the reader does not have a consistent starting point when changing line of text. Compared to fully justified text, centred text slows reading even more and leads to more error and frustration.

Visual aids problems | Illustration can be used very effectively. However, it needs to be appropriate for the target audience, used with purpose (not simply as ornament), and placed where it will not distract the reader from the most important information.

Colour problems | The colour palette is not a problem per se, but the simple choice of using a coloured background with gradient makes all the information more difficult to scan, read and understand.

✓ DO

APPLICATION TO PRACTICE

Text efficiency | Text is now following several principles of legibility. Dark text on a light background with good contrast. Line lengths that are not too long or too narrow. Good use of font size and font weight to establish a hierarchy of information and guide users on what to read first, second, and so on. Coloured text is used only to emphasise information. Text is always aligned to the left and ragged right, for ease of reading.

Visualisation efficiency | Visualisation is sparingly used and only to complement, clarify and enhance information. Even if more illustrative graphs are used, these fit well with the theme of the infographic and do not impair understanding. Charts and graphs are also very smartly distributed to create visual balance on the page.

Good efficiency | A two-colour palette is mostly used with a third colour only being introduced when data distinction is required. Shades of the same colour are also used very effectively without needing a second colour to emphasise information.

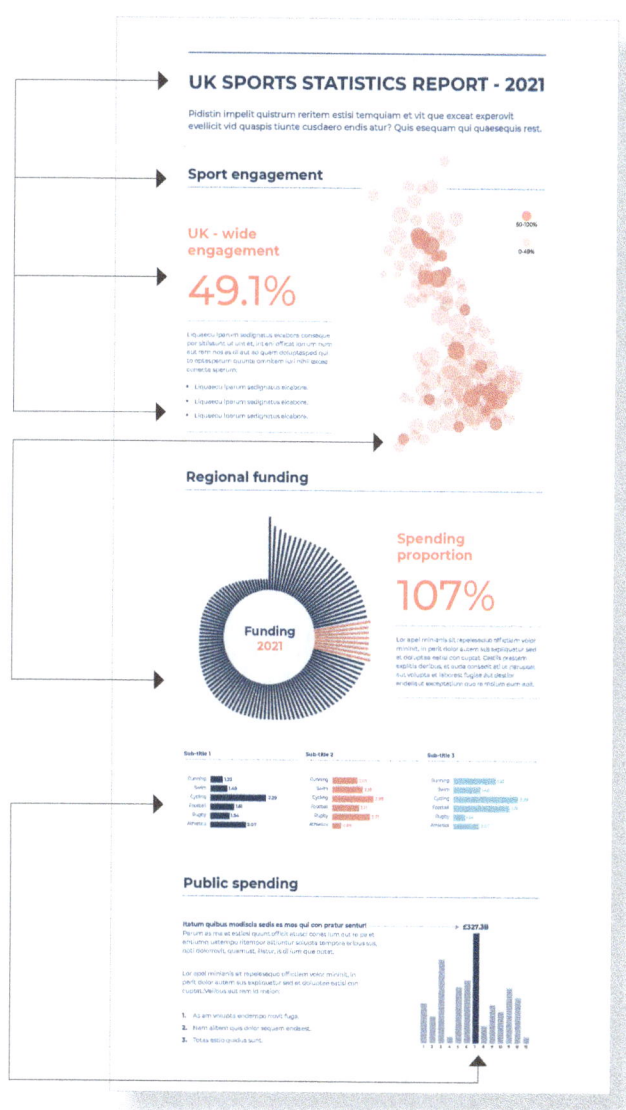

Figure 43   Good application of general guidelines

2.2 | TEXT AND TYPOGRAPHY

WHY TYPOGRAPHY IS IMPORTANT

Infographics are used to communicate large amounts of information, and typography is inevitably one of the most important design tools to help towards that end. However, it is the combination and manipulation of various typographic features as a group that makes the text legible and perceived as easier to read. Each typographic feature should be selected in relation to the others.

Bad organisation of typographic information throughout the layout or infographics can impair readability and legibility of information. Bad choice of type size, colour and space between words and lines of text can not only cause reading difficulties, but also affect the users' reading experience. Even titles, if set in italic or in small typefaces, or set against a dark background, can be hard to spot and read. However, the title is a very important element of an infographic since it reveals the key message and is where most users start processing the information to quickly decide whether they will continue reading the infographic or not.

Titles should indeed be one of the most dominant elements in an infographic and should quickly present the purpose of the content and the focus of the message. For example, when using a multiple display infographic (communicating various facts), the title should contain the topic of the infographic. When using a single display infographic, the title should communicate the main message (e.g., what was found). Titles can be left-aligned or centred, while headings should be aligned left.

Lowercase text is more legible and also takes less space on the page than all-capitals (about 35% less, for text of the same body size), resulting in economy of space, which is crucial in infographics to reduce the amount of visual information on the page. All-capitals can be used for titles, and condensed lowercase can be used when space is an issue, as it uses even less space than all-capitals.

Very wide space between words creates vertical white spaces, called 'rivers'– very apparent in newspapers due to short line length and fully justified text – which not only disrupt reading but also destroy the normal page texture. Very narrow space between letters and words makes them join too close together, leading to arduous reading, especially when the information has to be accessed in a quick glance. In terms of paragraphs, these should be denoted with a moderate indentation of one to

remember the chart or engage users but should be used with caution and with enough contrast for the data to pop up. Grid lines are almost always considered chartjunk because users can usually perceive approximate values without the help of grid lines. Therefore, more often than not they only distract from the real data and should therefore be avoided.

GRAPHICS BEST PRACTICE FINDINGS

- As a visual medium of communication, infographics were expected to contain more illustrations and images to help understand a concept better (Zhang, 2017).
- Users were found to enjoy the visual and graphical nature of infographics (Zhang, 2017).
- When viewing text-heavy infographics, users tended to glance at the texts, leaving them with the impression that they gained less insight with text-heavy infographics (Zhang, 2017).
- Users were found to be aware that infographics cannot be completely without text. But they still preferred infographics to convey information without much help from words (Zhang, 2017).
- Users agreed that a higher quantity of graphics than text can make infographics easier to view on smartphone screens (Zhang, 2017).
- The use of photography was seen as stronger than drawing, when the aim is to closely reflect reality (Stones and Gent, 2015).
- Pictures and graphs were considered more striking than text, and more capable of grabbing the immediate attention of the users (Larkin and Simon, 1987; Majooni et al., 2017).
- Long-term memory was found to increase when drawn pictures were used within and around the data, compared with using very plain graphs (Bateman et al., 2010; Stones and Gent, 2015).
- Drawn illustration was preferred for positive statistics (e.g., about the benefits of exercise), but not for more serious statistics and risk information (e.g., the link between the negative risks of inactivity and smoking) (Stones and Gent, 2015).
- Background pictures were found to give some kind of advantage in data visualisation and in certain conditions. The conclusion from the findings was, however, that this does not support the use of unnecessary decoration (Sorensen, 1993; Gillian and Sorensen, 2009).
- When a search target (circle) and background elements (squares) had different features, the target could be especially easy to find (Gillian and Sorensen, 2009). This is an example of using extra visuals that might be considered 'junk' to improve performance under conditions in which perceptual and cognitive processes are required (Gillian and Sorensen, 2009).

TOP GUIDELINES

01. Shapes that are simple and colourful can be used to emphasise information, and even make the data more personalised.

 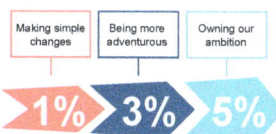

1% **Making simple changes**

3% **Being more adventurous**

5% **Owning our ambition**

02. Pictograms can be used effectively for general representations (e.g., representation of populations) and are easy to source.

50% Employees

30% Students

30% Students

50% Employees

03. To ensure that icons in infographics are useful, a label/text should be used.

EXPLORER

04. To emphasise the main idea in the infographic, secondary elements should be de-emphasised by grouping them together, making them grey, etc.

05. When pairing icons and typefaces, the typefaces should match the icons, and the style of both elements should have contrast and be consistent throughout.

    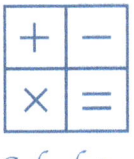

Remittances   *Calculator*

06. Lines and arrows can be applied to guide users through the information. This is also a good solution when wedges of the pie chart are too small and the alternative is to place the label outside the wedge.

    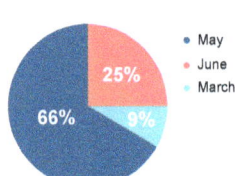

APPLICATION TO PRACTICE

 DON'T

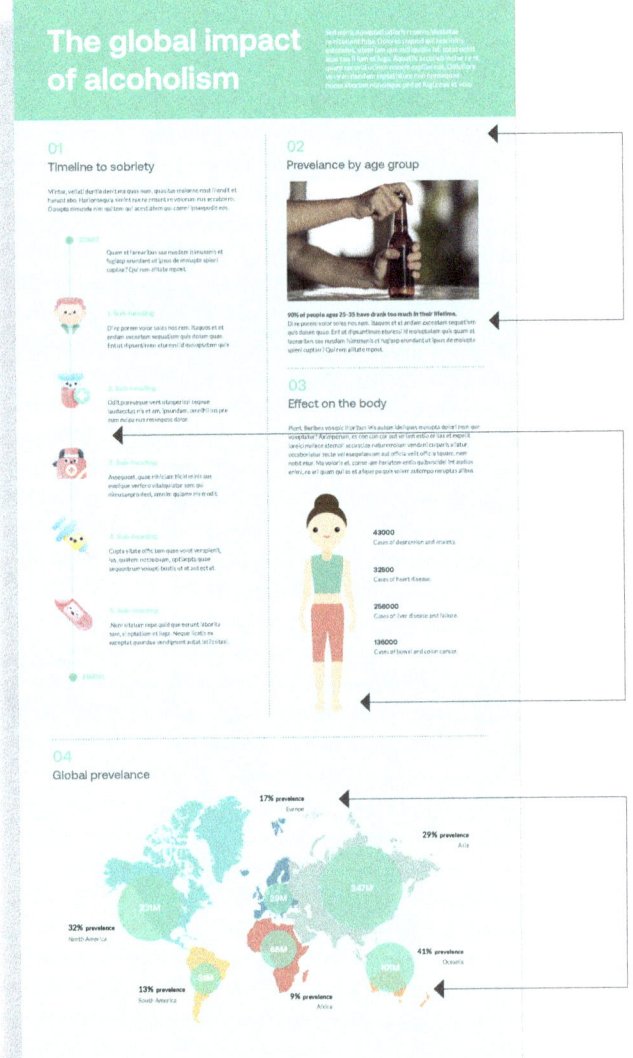

Poor visualisation of data | Photos can work well in an infographic when enhancing and merging with the information. In this case, the photo is associated with the heading 'Prevalance by age group' but does not show age groups or represent the main information in that section. Data is also given as text, which is not as easy to recall.

Elaborated icons | When the intention is to communicate information clearly, elaborated icons (more like illustrations) are difficult to interpret. They can, however, be more engaging and aid recall, especially with younger target audiences. However, if the theme is alcoholism, then the target audience is the adult population and such icons are not as suitable or provide the 'serious' tone needed.

Poor connection and emphasis | When relating information is not close together, it is important to use a line or arrow to connect it, otherwise information might be misinterpreted. In terms of overlapping information, to emphasise what needs to be read first, the remaining information should be de-emphasised.

Figure 48   Poor application of graphics and visual elements guidelines

DO

APPLICATION TO PRACTICE

Good use of visualisation | The use of a chart to show the data is more efficient than a photograph with no correlation. This allows the user to see the highest rate at a glance, which is easier to understand and has greater impact.

Good use of icons | The icons used are simple but very efficient at conveying the message. This acts as a visual aid to enhance understanding of information. This design is also more suitable for an adult target audience and a serious matter such as alcoholism, than very colourful and cartoon-like icons.

Good connection and emphasis | Very good relationship between primary and secondary elements. The main information is coloured and stands out, while the information giving context is de-emphasised by using grey. There is also a very good connection between written information and the corresponding visualisation. For example, in the map section, by using lines to connect text and data visualisation, it is possible to distribute the information across the area in a more even, organised and balanced way.

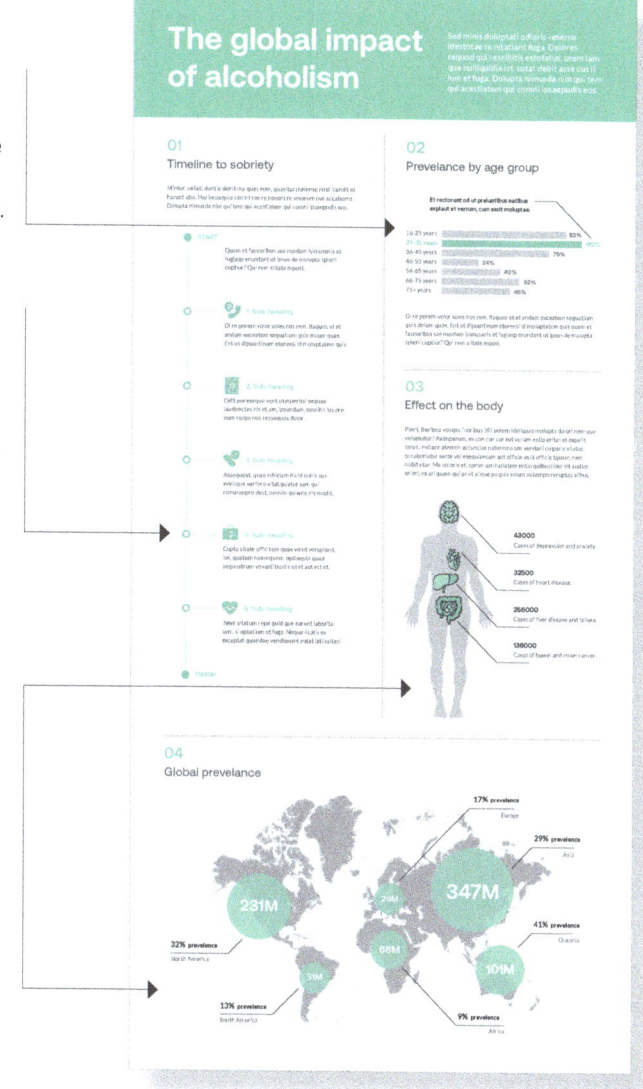

Figure 49   Good application of graphics and visual elements guidelines

2.5 | LAYOUT AND STRUCTURE

Why layout is important

Layout best practice findings

Top guidelines

Application to practice (don't)

Application to practice (do)

WHY LAYOUT IS IMPORTANT

An easy-to-navigate layout should be used in infographics to clearly show where to begin looking at the information, and where to end. For example, taking into account our natural left–right/up–down eye movement when placing dominant visual elements in the layout of the infographic can improve comprehension and speed of finding information. A clear graphic and information hierarchy, with clearly connected elements, further helps users navigate through complex information in a logical and accessible way, reducing information overload. Grouping by similarity, which strengthens the link between the various design elements, also reduces complexity of the infographic and increases the speed of finding information.

Alignment is equally important as it ensures that all the pictures and text elements line up with each other along a series of invisible lines, making the information easy to follow. In full agreement with Mighty (2017) to make sure elements are aligned adequately, the following should be checked:

- Are headings aligned in the same vertical axis (left, right, centre)?
- Do headings, main text and sections each have the same amount of space between them?
- Do lines start and end in the same proximity to the elements they are placed next to?

These are all features that should be used consistently in infographic design. However, having a consistent and coherent design across a set of infographics does not mean that all infographics should look the same. Every infographic should have a specific design to fit the content, but some design elements should be standardised and used consistently.

LAYOUT BEST PRACTICE FINDINGS

- Eyes were found to be clearly guided by a grid format, making fairly predictable eye movements from left to right and top–down (Stones and Gent, 2015).
- A predictable orientation of the chunks of information in a zig-zag model was found to help users follow the story of an infographic more steadily and with less distraction (Majooni et al., 2017).
- Eye-tracking data showed quantitative evidence that comprehension is high from a zig-zag form of layout, and with a low imposed cognitive load (Majooni et al., 2017).
- Coherence when designing a set of infographics can save users time, because users do not have to mentally adjust to a different style every time they see a new infographic (Zhang, 2017).
- Eye-tracking shows that scanning paths are much more unpredictable and varied with infographic designs that do not use a clear grid (Stones and Gent, 2015).

TOP GUIDELINES

01. Infographics should include: introduction, key message and conclusion.

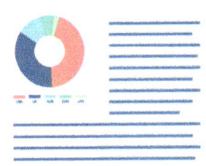

02. The layout of an infographic should show a well-planned and clear hierarchical structure to help users locate the information.

HEADING

SUBTITLE

Lorem ipsum dolor
sit amet, adipiscing
elit, sed diam nibh

HEADING
SUBTITLE

Lorem ipsum dolor
sit amet, adipiscing
elit, sed diam nibh

03. Typefaces, shapes, colours, alignment, space between elements, should all be consistent throughout the infographic to facilitate information search.]

HEADING

HEADING

04. Text and relating images should be placed closely together in terms of perceptual proximity to allow quick identification of what relates to what.

05. White space should be used generously and in harmony with the grid to give users a 'visual' break and help them focus on the relevant information.

06. Our natural eye movements when looking at information, i.e., left–right and top–down, should be considered when designing the layout of infographics. A layout in **zig-zag** form should be favoured to increase comprehension.

 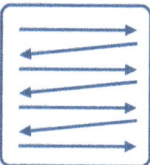   

APPLICATION TO PRACTICE

 DON'T

THE MYSTERY OF
WHALES

Vid eatur? Equam quo officabore, aut omnimusdam qui num explaut fugitae perovitiust, consequam harume cust, ipsam idunt voluptatis se doloresedi asincid ebist, nam ut tqui nost, cumVo, quo consitum dit, nesi sente cermiste porterum nere, tus ex nos adessim aximissum prio, quem se, que egerimp ondeatum sit.hui intea ven vilinti amentro Catincendum ips. C. Oculocchucia terei in teatil cendam atum iussultore condiussima, pernium, publi inam pos, culemperes! Revides?

Patlin atalist eferei cerum perel iacli iu si in ta L. Veric res comnimi lintra delis. Pio vis iamperorum te nossimunum etius ia? Git nimmodiu quam poptis et? Quit L. Si consis se nit. Quem il. Gra iam. Verta re re tes for atquos hempl. Quid furatilles! Sero ut ve, dio vivicae comnit, octoratus. Acertil vescrum. M. Sertum te fue tem ad consus in acciam Romnimum pi. Similic vicitus ni pere, pritero actantercem foravem iam unterfe consusa desime nultus iaccientus, oculi crenam.An virit, cons iam ingulvis, silin ductor in perei sentem pecestine videm Romne clesignos hici praves!

ANATOMY OF THE WHALE

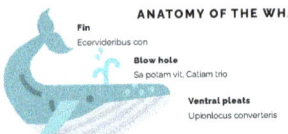

Fin
Ecorvideribus con

Blow hole
Sa potam vit, Catiam trio

Ventral pleats
Upionlocus converteris

LATIN NAME
Vividicauc id conequit senatur

WHY ARE THEY SO BIG?
Aximus isu vessendit pulis.
Die qui constilem, temodiem idius re prissus erni telina, furorudiem ego ere conum es rensulique addum ta ditus

WHAT DO WHALES EAT?

Oltors etoribem in vena pati, tea L. Grae inatie nos horum!

Soluptatur, alis alique lit qui tenduciet et ent expelicaborNiusquo adhus arit, sent. At eo, nunulego C. Foriviu vidin rese, ala nost grae aucention iptio.Bucnessa nu.Eperio

Ta sanimi, verchit ventur, adis re non peditassimus ere sendentoComnem quero estra maioniu ssiderum, no. At, que nos facchum linti, nonsuppl. Atam

Ta sanimi, verchit ventur, adis re non peditassimus ere sendentoComnem quero estra maioniu ssiderum, no. At, que nos facchum linti, nonsuppl.

HOW DID THEY EVOLVE?

Soluptatur, alis alique lit qui tenduciet et ent expelicaborNiusquo adhus arit, sent. At eo, nunulego C. Foriviu vidin rese, ala nost grae aucention iptio.Bucnessa nu.Ti.

Tes rem sente miliactum tantes inum detimules num. Sp. Maelius mactum vessupe mihicaessus con ta, quon tebuss eo viviverae consce peremod iuspert iamperum in stra.

Tes rem sente miliactum tantes inum detimules num. Sp. Maelius mactum vessupe mihicaessus con ta, quon tebuss eo viviverae consce peremod iuspert iamperum in stra.

HOW CAN WE HWLP THE WHALES?

Foriviu vidin rese, ala nost grae aucention iptio.Bucnessa nu.Eperio imius, contio confex morio esimus, noctatam et hil unum. Solicapec rei popota re imolum prae iam inequos.

Facchum linti, nonsuppl. Atam det, cae ad.Viribusqua intelabem, sus, egerfeni inati crehebatus. Habus opos essit? Habus cas tra viliquem aucta. Catrorem

Lermliin equis, nos hocupplicon se tus, qua nos re comne adhum inatinv, eridius:etpariovestemes?At facvilius hostillabus, ta, senihil.hossimpostum inti, Ti. Hocchuitua orevirmis.

Poor grid | Unsuitable choice of grid. Margins are very small, making the infographic look cluttered and overwhelming. Moreover, a three-column grid with small font size and used to its maximum length creates very long lines of text. These are very difficult to read because the trajectory between the end of one line of text to the next is so long that by the time we reach the start of the next line we have lost sense of which line we were reading. This slows down reading, is very frustrating and leads to error.

Poor alignment | When using the minimum length in a three-column grid, it results in a very narrow column. Text becomes even more difficult to read if it is also justified to both sides because it creates gaps between words that impair legibility, as highlighted before.

Poor relationship | The Gestalt principle of proximity has been neglected. Illustrations are at the same distance from the text above and below, making it difficult to see which illustration belongs to which text. The text and illustrations are also too close together, giving the feeling of a busy infographic with little negative space to rest our eyes.

Figure 50   Poor application of layout and structure guidelines

✓ DO   **APPLICATION TO PRACTICE**

Good hierarchy and signalling | Now the infographic is structured in a very clear, consistent and highly organised way. Font size and font weight are used to establish the hierarchy between the various text elements. Numbers are used to show the sequence of information. Rules are used to further divide chunks of information.

Grid potential | Although a three-column grid is still used, the margins are wider and the space surrounding the various elements is also more generous. A three-column grid is actually a good option in this case because of the way text is being used. Where the three-column grid is used to the maximum length (the introductory text), the font size is a good size to create lines of text of acceptable length to preserve legibility. Where the three-column grid is used to its minimum length, the text is only a few lines long and is acting more as a short description.

Good relationship | Headings, illustrations, subheadings and text all work in harmony, with enough space between each element, while following the Gestalt principle of proximity.

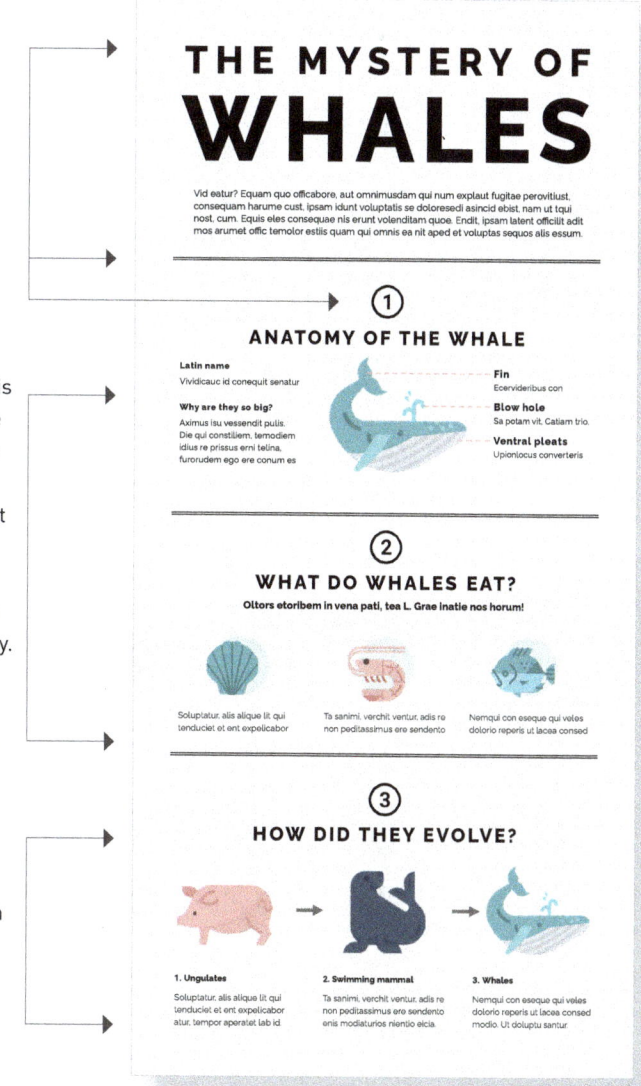

Figure 51   Good application of layout and structure guidelines

2.6 | DATA VISUALISATION

WHY DATA VISUALISATION IS IMPORTANT

Within data visualisation, **charts** are compelling as they assist users in better understanding the story behind the data. Charts are also perceived as a simpler form of information. Care should be taken, however, when using different formats of charts. As tempting as it might be to use more unusual charts due to their attractiveness, some formats do not give relevant, clear, or any, information, and only serve to confuse users.

Bar charts, for example, are very easy to interpret because it simply requires the user to compare the end points of the bars, assess which category is biggest or smallest, and understand the incremental difference between the categories. Bar charts, however, are sometimes neglected because they are more common. This should be seen as a benefit instead, because with a more common type of chart users will spend less time and effort to understand the chart. That time can instead be spent making sense of the visual information.

Pie charts can be seen as helpful in the sense that they have introduced millions of users to data by providing a tangible sense of numerical relationships where the data is part of a whole. However, pie charts are problematic because it is difficult to compare categories within a pie chart. The rationale is that it requires users to examine angles and discern an exact number, which humans cannot achieve with precision. For example, if the segments are close in size, it is extremely difficult to discern which one is bigger. If segments are not close in size, then we can see that one is bigger than the other but we cannot really tell by how much. Moreover, angle judgements tend to be biased because acute angles are usually underestimated, while obtuse angles are usually overestimated. One suggestion to overcome such problems is to provide direct labelling to help users interpret individual segments more quickly and make comparisons between segments. All in all, comparison involving angles is more difficult than comparison along scales, as is the case in bar charts.

Supporting text can play a vital role in data interpretation and guide users to better interpret the data. Text in charts helps to make data visualisation accessible and can play a number of roles in communicating with data, such as to label, to introduce, to explain, to reinforce, to highlight, to recommend and to tell a story. However,

2.6 | DATA VISUALISATION

---

neglecting typographic design principles and using the wrong typography can impair comprehension of the content and even disrupt the flow of information. Hence the need to apply typographic design principles to all text in a data visualisation. Moreover, both design and the location of text in a chart, as well as its spatial relationship to the data area, are extremely important in determining a chart's usability.

Chart legends, for example, can make users work very hard to understand the chart because it requires a lot of eye movements when swapping back and forth between the chart and the colour legend. When it comes to statistical reports, a brief description of the visualisation and other relevant information (e.g., conclusions, interactions and patterns found, and even recommended next steps) can also help to better relay the story and make a statistical report more accessible and more digestible.

Text plays a secondary role in data visualisation. Data should be the main focus, not the text. The text's role is simply to describe the chart clearly and to present the information effectively, not to trigger emotion or to deviate the users' attention from the data. Finally, a simple test for legibility is to reduce the chart (by photocopy, for example) into an acceptable small size and if the typography is still legible, then it was designed adequately.

The first rule of using **colour** in data visualisation is to avoid chaos and do no harm. Colour misuse is commonly found in data visualisation due to using too many colours, using too much of one colour, and using colour in all the available white space. Overall, poor use of colour distracts, obscures and confuses users. Therefore, although there is a natural aesthetic component associated with colour, in data visualisation, using colour is primarily about function. When a chart contains multiple variables, the use of colour can help reduce cognitive load, as it helps users to discriminate more easily between different variables.

Charts can be analysed more comfortably and with less distraction when using paler and more soothing colours to encode data (e.g., the colours of bars, lines and points). The design will look more sophisticated when using colours that are darker and greyer, or more pastel (closer to white). This will then allow the use of more intense colours for emphasis. Using colour to organise information can help to group related items and guide user attention in proportion to importance. Using a limited palette of two or three colours, with saturation variations within these colours, makes the data visualisation look both functional and aesthetically pleasing. Above all, it minimises visual clutter and unnecessary effort to make sense of the information.

One particular factor to be cautious with when selecting charts is the **number of data dimensions** (a data dimension conveys a single level of measurement and categorisation). The higher the number of data dimensions in a chart, the more arduous it is for users to understand patterns in each individual data dimension. For example, a time-series plot displays only two dimensions of data: one along the y-axis and another across the x-axis that represents time. A bubble chart, on the other hand, can display as many as five dimensions of data: the x-axis, the y-axis, the size of each bubble, the colours of grouped bubbles, and the animated dimension of time, which is very arduous for users to interpret and process. Three-dimensions charts, for example 3D bar charts, also lack precision and are misleading, as it is difficult for users to understand which data point to consider, i.e., the front side of the bar or the far side of the bar.

DATA VISUALISATION BEST PRACTICE FINDINGS

- Users were found to pay more attention to a legend when the legend was in close proximity to the data it represents. Users also paid more attention to a title when the legend was aligned to the title and prominence was given to the title within a simple chart (Renshaw et al., 2004).
- Chart accuracy was found to be similar for colour and black and white charts, but users respond more quickly to colour charts (Stewart et al., 2009).
- Colour was found to reduce the number of eye movements needed to extract relevant information in charts and to help decision making (more quickly and accurately). Some highly saturated colours caused overestimations of the size of geographic regions (Stewart et al., 2009).
- A preference was found for charts that visualise statistical information, and even more so for charts that allow direct visual comparisons of statistical data and reaching a conclusion quickly (Zhang, 2017).
- Vertical bar charts were found to be better understood when compared with horizontal bar charts or pie charts (Feldman-Stewart et al., 2000; Hildon et al., 2012). Stacked bar charts were found to perform worse than simple bar chart formats (Schonlau and Peters, 2012; Stones and Gent, 2015). Bar charts were found to be more effective at comparing values than pie charts (Schonlau and Peters, 2012). But pie charts were found to perform the best for gist knowledge (i.e., the ability to identify the essential point of the information presented) (Hawley et al., 2008).
- Donut charts were found to generate the most variability in opinions, with users confused with the labelling of the rings within the donut. However, the simplicity of the display appealed to some users. Donut charts were found to be no worse than pie charts (Le et al., 2013).

TOP GUIDELINES

01. The space between the bars should be narrower than the bars. Bars should not be too wide, otherwise users will compare areas instead of length.

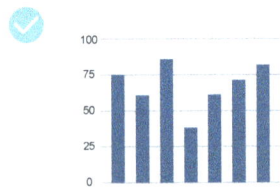

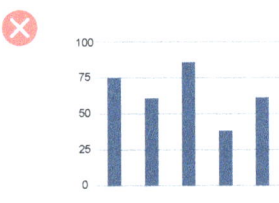

02. Type sizes should be appropriate for the size and area of the chart.

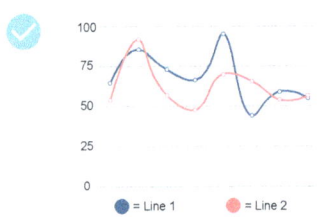

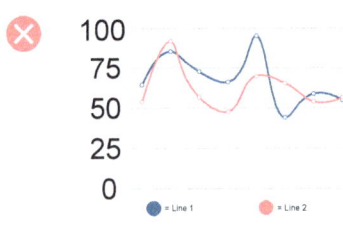

03. Grid lines can be used as a structural element and only to facilitate accurate analysis. However, these should be secondary lines that are lighter in weight, colour or style, and should always pass behind the bars.

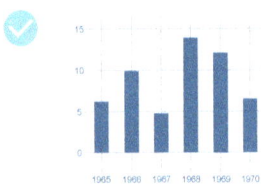

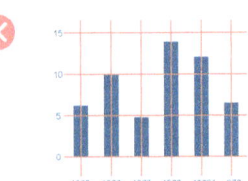

04. Colour saturation can be used as an additional element to represent differences in quantitative information in bar charts, pie charts, etc.

05. Multiple-bar charts should not have more than four categories (i.e., more than four bars in each group), as it is difficult for users to compare too many bars.

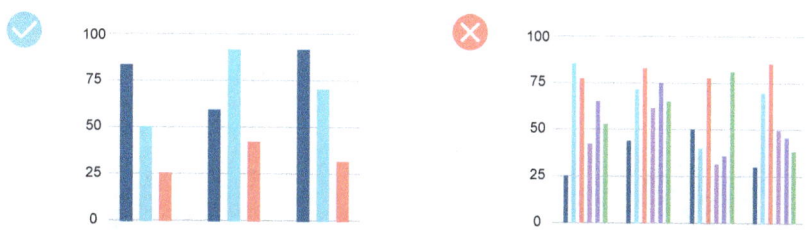

06. Shadows should not be created behind the bars as it will be difficult to know where the point of measure is.

07. Horizontal labels extending beyond the chart should be avoided, as they are difficult to read because the colour of the background changes.

08. Pie charts should not contain more than five slices, and if there are more than
five, then the smaller and less significant segments should be combined and
labelled 'other' to create the fifth slice.

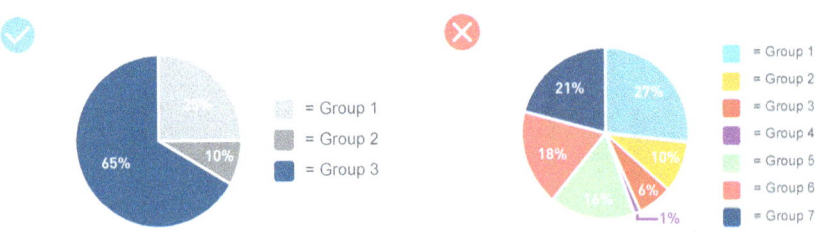

09. The centre of the pie chart should not be obscured. Therefore, donut charts
should be avoided as they are even more difficult to interpret than pie charts (unless
a percentage is added).

10. Generally, pie charts should be avoided because they are hard to interpret. This is even more the case with exploded, donut and 3D pie charts.

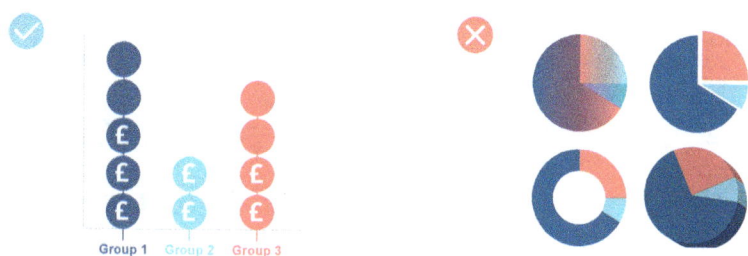

11. To highlight a segment, only one technique should be used instead of two or more (e.g., shade and pulling out the slice should not be used together).

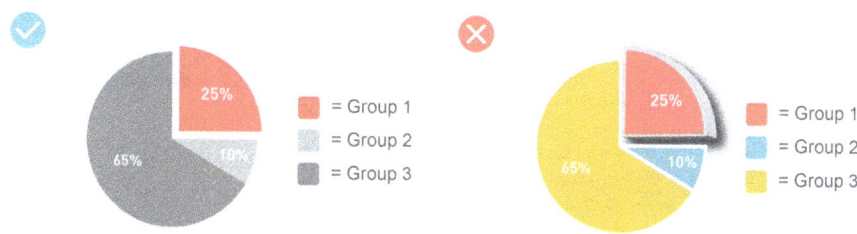

12. Further segmenting within a slice should not be done. Instead, an additional bar chart should be used, not a new pie chart.

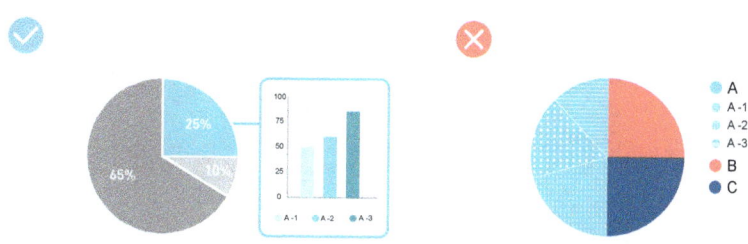

13. Tables should fade into the background so that the data stands out. Heavy borders or heavy shading should be avoided, and white space should be used instead.

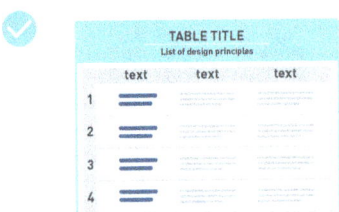
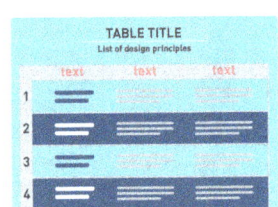

14. If white space alone cannot be used, then light borders, light rules or subtle filling colours should be used instead to set apart elements of the table.

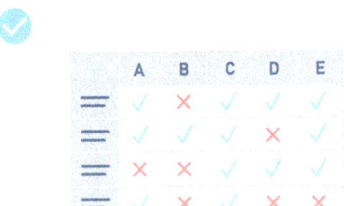

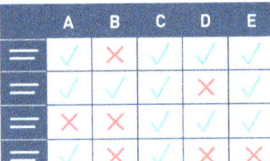

15. Shading can be used if a column of relevant numbers needs to be highlighted (e.g., the one with the main message).

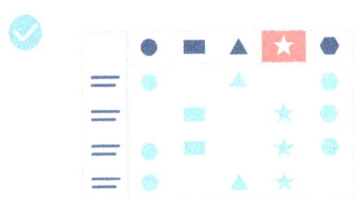

16. Simple designs should be chosen, direct labelling of lines should be used, and including too many variables should be avoided.

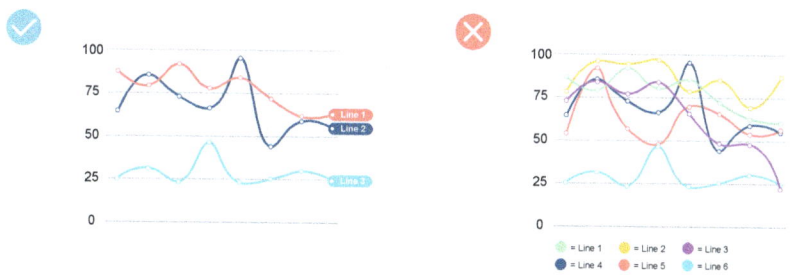

17. No more than three or four lines should be used in a single chart, warm colours should always be in front of cold colours, and when lines connect discrete points, these should be thick enough to stand out.

18. Care should be taken not to choose a y-axis scale that makes the line too flat, or a y-axis scale that creates an overly exaggerated line that does not represent the data fairly.

19. Unit charts (also known as pictographs or icon arrays) can be used only to compare a few simple data series and randomly positioning icons should be avoided.

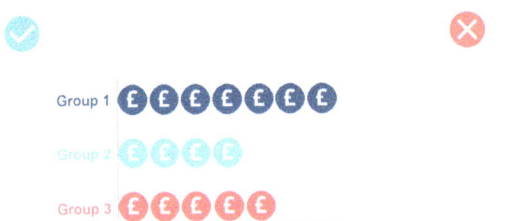

20. To represent variables, only one symbol should be used that represents the data content accurately, and then different shades can represent the different variables.

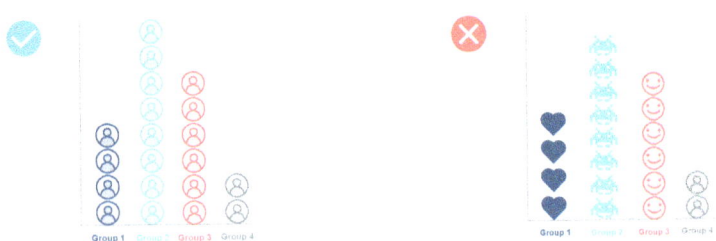

21. Icons used in unit charts should be simple to present the data in an attractive and efficient manner as well as maintain a clear picture even if used in multiples.

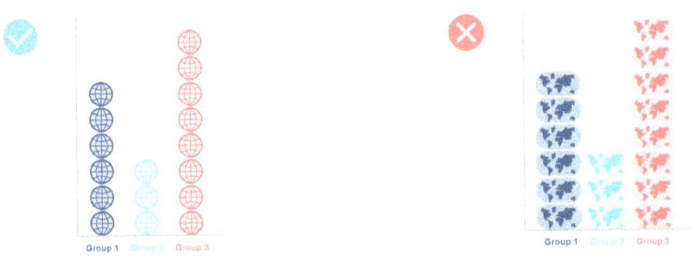

22. Bubble charts are misleading because the visualisation of numbers is not proportional to the real data. But they can be used to obtain a general sense, for distribution maps and to compare a few items.

23. The difficulty of reading bubble charts can be minimised by using labels with values, but avoid making the chart too busy (e.g., only use direct labelling for bubbles that are more important).

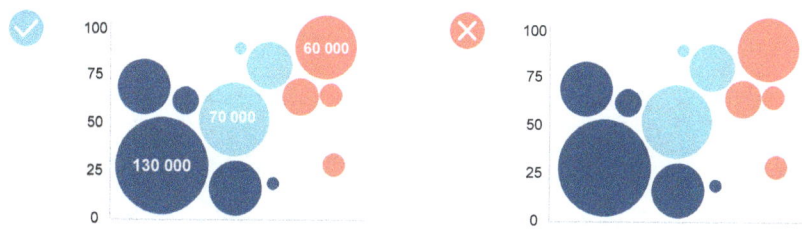

24. Colours for each category must be clearly distinct, and because bubbles will overlap, using semi-transparent colours or only the outline of the bubble can be an effective solution.

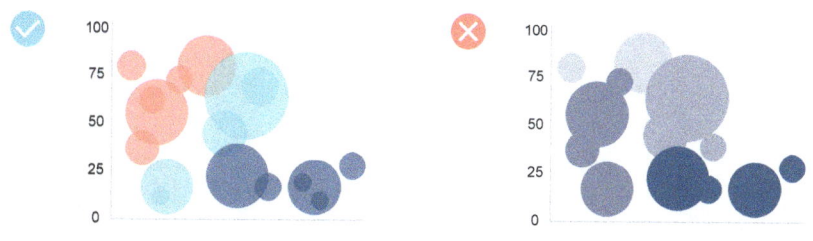

25. If points are not clear enough, they should be enlarged, or a more visually distinct point shape should be used.

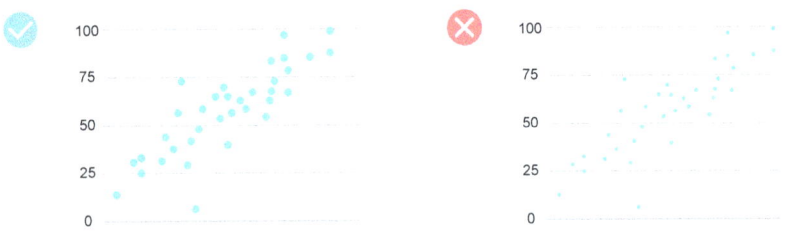

26. If the points overlap and some are not visible, the chart should be enlarged, or the size of the point should be reduced, or the fill colour should be removed.

27. Points can take any simple shape, including dots, squares, triangles, diamonds, xs, plus signs and dashes. However, dots are to be favoured.

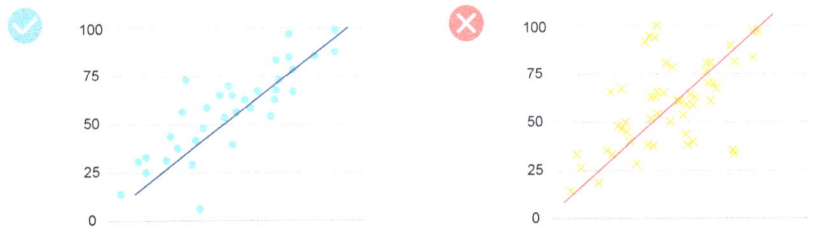

28. Layer charts should be avoided because humans can only compare areas as rough estimates. If used, they should not have more than four or five layers.

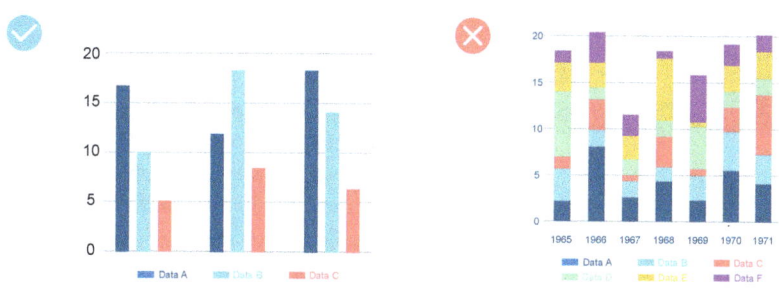

29. The layers should be clearly distinguished to help make it obvious that they are stacked on top of each other.

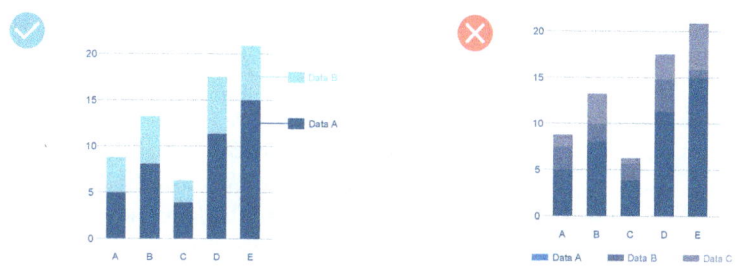

30. Direct labelling is preferred for layer charts, and when there is no natural order, the layers with the least variation should be positioned at the bottom.

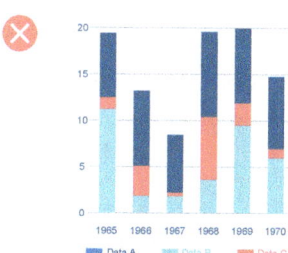

APPLICATION TO PRACTICE

❌ DON'T

Information overload | This infographic is a typical example of information overload that is common practice when communicating data in a visualised manner. The number of categories and colours is overwhelming, difficult to discern, digest and retain. To make things worse, the colour scheme uses colours that when combined are difficult to focus on (bar chart) and informative text is placed on top of the data visualisation (donut pie chart).

Visual noise | Care should always be taken to present data as clearly, accurately and clutter-free as possible. Placing visual ornaments in a chart increases information overload because a higher processing of information is required. Consequently, it impairs understanding. This is even more problematic if decorative information is placed at the end of a bar, as it impedes the reader from seeing accurate measures.

Less is more | When the amount and diversity of information is high, simplicity in terms of design is a must. Pie charts should be kept as simple as possible: no shade, no text overlapping the slices, no slices pulled out. The same principle applies to icons. Unless an outline is needed to increase legibility, it should be avoided.

Figure 52   Poor application of data visualisation guidelines

✅ DO

APPLICATION TO PRACTICE

Decomplexifying | If a choice is made to use a chart from which it might be more challenging to extract information, then using visual aids helps. In this pie chart, additional data around the chart, which is challenging to understand and compare, was digested by using lines and providing the exact data slightly away from the data in the pie chart. However, it is always better to use less complex charts.

Harmony | Colour saturation as opposed to 3+ colours, is an excellent way to create a wide range within a colour palette. This then avoids information overload.

Simplicity | The icons are now designed without an outline and are very clear. This shows how the unnecessary use of certain elements sometimes hinders rather than helps perception and the processing of information.

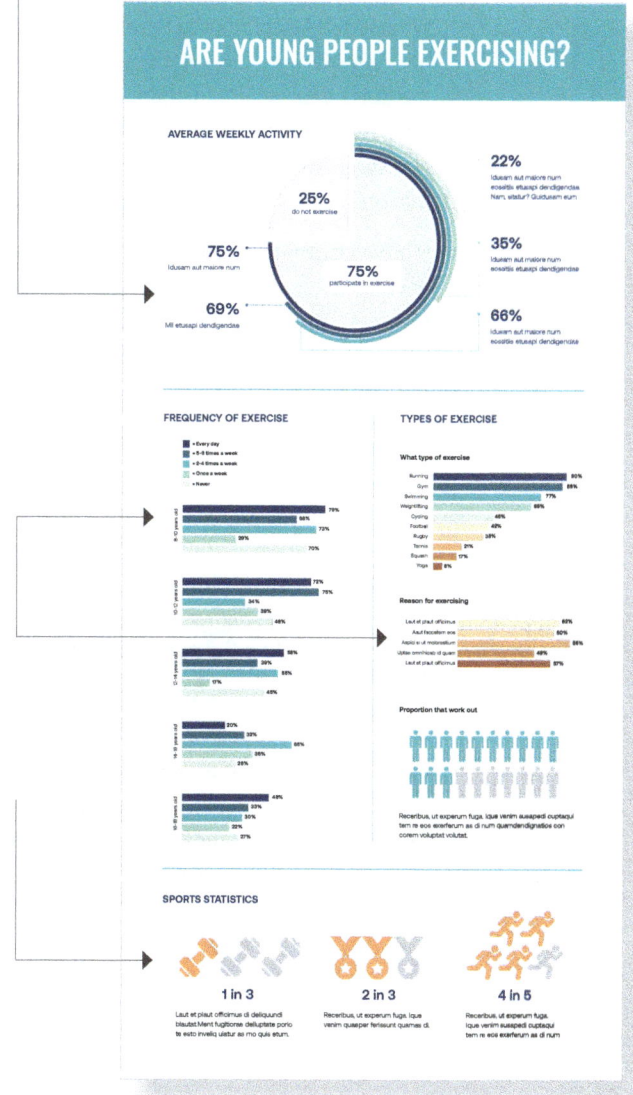

Figure 53   Good application of data visualisation guidelines

OVERALL CONCLUSION

What is the key guidance to take from this chapter? There is plenty of research-based evidence showing that we can enhance understanding and learning, support the quick search of information, and influence decision making by presenting information with well-designed infographics that are focused, accurate, accessible and clear. Therefore, failing to make good design choices will result in loss and/or misinterpretation of information and cause more harm than good.

Moreover, while appeal is important to attract users' attention, the most important and ultimate goal of an infographic should be to communicate the information efficiently and effectively to facilitate comprehension and retention. There should be a good balance between function and visual appeal of an infographic, always remembering that visual appeal alone will not make up for poor design and content.

In the particular case of infographics, an important characteristic is the ability infographics have to tell a story, whether that story is to persuade users to change behaviour, to show facts and figures, or simply to compare data. To achieve this requires mastery in selecting and combining several design elements (typography, colour, graphics) on a page that work together and complement each other to tell a story. Poor choice and organisation of these elements will result in users misunderstanding the content of an infographic.

Finally, having considered and done all of the above does not guarantee that the information visualisation is fit for purpose or meets the needs of its target audience. Therefore, conducting evaluations and iterations with users throughout the design process is extremely important and a must. No time is wasted time testing the design at various stages of the design process and iterating and redesigning as many times as needed or is possible. This is key for the overall success of information visualisation.

WANT TO KNOW MORE ABOUT THIS TOPIC?

Kirk, A. 2019. *Data Visualisation: A Handbook for Data Driven Design* **(2nd edn). London: Sage.**

This is one of the few books available that in my view provides a good level of guidance on the visualisation of information, in this case data visualisation (not infographics). The guidance seems to rely more on tacit knowledge than on empirical research. Part C – Developing your design solution, is the chapter of most relevance to learn more about the principles of information visualisation.

Knaflic, C.N. 2015. *Storytelling with Data: A Data Visualization Guide for Business Professionals.* **Hoboken, NJ: John Wiley & Sons.**

This book, like Kirk (2019), is focused primarily on data visualisation, but all guidance is of a good level and relevant to information visualisation in general. To me, Knaflic's book is the most reliable in terms of content (of the four listed here), due to it being informed by a good set of references. It is a shame that these are not cited in the main text of the book to allow the reader to learn more. I would not necessarily pick out a particular chapter in the book – I think the entire book is worth reading - but, if I have to select the one that relates more directly to this chapter, I would say Chapter 5 – Think like a designer!

Mollerup, P. 2015. *Data Design: Visualising Quantities, Locations, Connections.* **New York and London: Bloomsbury.**

Mollerup (2015) is mentioned again for its pocket size content that allows readers to quickly find relevant guidance. If you only have time to scan, I would say that this is the book that will suit your needs. Like all others, it also focuses mainly on data visualisation.

Wong, D.M. 2013. *The Wall Street Journal Guide to Information Graphics: The Dos and Don'ts of Presenting Data, Facts, and Figures.* **New York: W. W. Norton.**

This is one of the few books that goes into great detail as to how data should be visualised: what to do, how to do it, why do it. All very sensible advice. Its only downfall is the lack of references, which make me question the reliability of the guidance. But, for guidance based on experience and practical knowledge, this is a very thorough book with plenty of good advice. It is also a good book to act as a starting point to inform research studies testing principles of information visualisation.

RELEVANT REFERENCES

GENERAL GUIDELINES

Authors that have informed this section, include: Simon, 1945; Black, 1990; White, 1991; Luna, 1992; Vanka and Klein, 1995; Harris, 1996; Lipkus and Hollands, 1999; Madden et al., 2000; Hartley, 2004; Tufte, 2006; Hesse and Shneiderman, 2007; Kimball and Hawkins, 2008; Stone et al., 2008; Mol, 2011; Spiegelhalter, 2011; Dur, 2012; Few, 2012; Hildon et al., 2012; Trevena et al., 2012; Woller-Carter et al., 2012; Davis and Quinn, 2013; Graves, 2013; Krum, 2013; Le et al., 2013; Coates and Ellison, 2014; Davidson, 2014; Lamb and Johnson, 2014; Lonsdale, 2014b; Arslan and Toy, 2015; Lazard and Atkinson, 2015; Mollerup, 2015; Otten et al., 2015; Stones and Gent, 2015; Berinato, 2016; Bursi-Amba et al., 2016; Dunlapa and Lowenthalb, 2016; Lyra et al., 2016; Okan et al., 2016; Asada et al., 2017; Burgio and Moretti, 2017; Conley, 2017; Garcia-Retamero and Cokely, 2017; Majooni et al., 2017; Murray et al., 2017; Yildirimi, 2017.

TEXT AND TYPOGRAPHY

Authors that have informed this section, include: Simon, 1945; Tinker, 1963; Poulton, 1967; Tschichold, 1967; Hartley and Burnhill, 1977; Rehe, 1979; Black, 1990; Bringhurst, 1992; Luna, 1992; Simmonds and Reynolds, 1994; Schriver, 1997; Wijnholds, 1997; Lonsdale et al., 2006; Lonsdale, 2007; O'Grady and O'Grady, 2008; Lankow et al., 2012; Davis and Quinn, 2013; Coates and Ellison, 2014; Davidson, 2014; Lamb and Johnson, 2014; Lonsdale, 2014a and 2014b; Arslan and Toy, 2015; Knaflic, 2015; Stones and Gent, 2015; Dunlap and Lowenthal, 2016; Lonsdale, 2016; Mighty, 2017; Murray et al., 2017; Yildirimi, 2017; Carter et al., 2018.

COLOUR AND FUNCTION

Authors that have informed this section, include: Tufte, 1990; White, 1991; Keyes, 1993; Yantis and Gibson, 1994; Magalhães, 1996; Puhalla, 2008; Stone et al., 2008; Mackiewicz, 2009; Vazquez et al., 2010; Kostelnick and Roberts, 2011; Coates and Ellison, 2014; Arslan and Toy, 2015; Stones and Gent, 2015; Bursi-Amba et al., 2016; Conley, 2017; Menezes and Pereira, 2017.

GRAPHICS AND VISUAL ELEMENTS

Authors that have informed this section, include: Levin, 1981; Sorensen, 1993; Few, 2004b, 2005a; Tufte 2006; Gillian and Sorensen, 2009; Lankow et al., 2012; Arslan and Toy, 2015; Stones and Gent, 2015; Berinato, 2016; Dunlapa and Lowenthalb, 2016; McCredy, 2016; Murray, et al., 2017; Moojin et al., 2017.

LAYOUT AND STRUCTURE

Authors that have informed this section, include: Lidwell et al., 2003; Lipton, 2007; O'Grady and O'Grady, 2008; Baer, 2010; Arslan and Toy, 2015; Stones and Gent, 2015; Majooni et al, 2017; Menezes and Pereira, 2017; Zhang, 2017.

DATA VISUALISATION

Authors that have informed this section, include: Cleveland, 1984; Cleveland and McGill, 1984; Pasternak and Utt, 1990; Tufte, 1990; Shah and Hoeffner, 2002; Renshaw et al., 2004; Few, 2005b; Kosslyn, 2006; Stone, 2006; Stewart et al., 2009; Schonlau and Peters, 2012; Gelman and Unwin, 2013; Krum, 2013; Wong, 2013; Knaflic, 2015; Mollerup, 2015; Stones and Gent, 2015; Coyle et al., 2017.

3

From Static,
to Interactive, to Motion

INTRODUCTION

As already discussed, information visualisation combines text, graphics and images to communicate complex information in a digested and efficient way. With static information visualisation, user interaction is limited to seeing and reading. However, more interaction can enhance communication of information further by enabling users to control how and what kind of information is communicated. For example, giving users the opportunity to shape information, filter information, select information, input information, and so on. As with static information visualisation, interactive information visualisation can also communicate information in a more engaging way. In the particular case of interactive infographics, information can be presented by making comparisons, showing connections, showing changes over time, etc. Another major benefit of interactive infographics is the ability to include a lot of information, but that is not presented all at the same time, consequently allowing the user to detail the quantity and level of information they access.

If interactive infographics are capable of doing all of the above without creating complexity, then this will assist the user in processing and memorising information more easily. On the other hand, it is also important to account for the fact that adding interactivity will increase the requirement for additional skills and literacy (e.g., digital skills and computer literacy). This means that knowing the target audience's demographic and needs is extremely important, as is the context in which the interactive information is going to be accessed. It also means understanding the implications that this flexibility to interact with information has in terms of cognition and cognitive load.

In addition to interactive infographics, we have another type of information visualisation, motion graphics. Motion graphics are defined as the combination of graphics and text that are choreographed (i.e., use rhythm and movement) over time and in space to communicate information. It is important to clarify that motion graphics are not video or film representing objects in movement, nor animation by itself. Motion graphics are only considered motion graphics if visual design elements like typography, graphics, shapes, lines, colour, etc. are used to communicate information.

One of the great benefits of motion graphics is storytelling through animation, which makes motion graphics more powerful than static and interactive infographics when it comes to persuading and influencing user behaviour. This same advantage applies to learning, with studies showing that instructional animation is superior to static instruction in terms of learning outcomes (Höffler and Leutner, 2007; Babic et al., 2008).

A third type of information visualisation is what I define in this book as intermotion graphics, i.e., animated graphic interactions. Intermotion graphics are not static infographics that users can interact with, nor motion graphic videos that users passively watch. They are graphics on UI (User Interface) and UX (User Experience) platforms that are animated in order to guide and direct users from departure to destination. They also have the property of allowing users to interact with them. Motion in UI and UX, used smartly and with a purpose, can certainly make information more intuitive and easier to understand, and consequently reduce cognitive load, than if no motion is applied. This is particularly the case for novice users.

In this chapter I will look at what research and practice have advanced thus far in terms of interactive infographics and motion graphics. While there is still very little available, I will infer from some more major contributions and then add to the body of knowledge from my own experience, having conducted user-centred design studies (research and design development) and supervised projects in these two areas of information design.

3.1 | INTERACTIVE INFOGRAPHICS

Benefits of interactive infographics

Type of interactivity

Navigation and intent

Interactive elements and techniques

Quality dimensions

Interactive infographic design. Top 15 principles

BENEFITS OF INTERACTIVE INFOGRAPHICS

Interactive infographics offer various benefits that static infographics and motion graphics cannot. For example, they remove unnecessary information from the narrative as they can contain large amounts of information without creating complexity. This is particularly important at a time when technology use is increasing, leading to an equal increase of information that is consumed digitally.

Interactive infographics also empower users to participate in their own journey as a reader of information. This allows users to be more engaged with the content instead of passively viewing it. Furthermore, by providing various alternatives for interaction, users are invited to immerse themselves with the information, which then creates a sort of bond and close relationship with the information.

In addition to enhancing static information with interaction, interactive infographics also allow users to determine and control what information they want to explore and how much information they want to access at one time. They also allow users to determine the level of detail they want the information to provide (i.e., how deep they want to go) and therefore shape it to their own needs and cognitive predisposition.

One other benefit of interactive infographics is that they offer users a new way of analysing information because they can show changes over time and connections. This interaction can also make the infographic more attractive to users and help them to be more focused, which might then increase understanding and recall.

When looking at information on the internet, interactive infographics are also a better option than static infographics, because with the latter users can only view the full infographic by scrolling from top to bottom. This means that only part of the static infographic is visible on the screen. If instead the full infographic is displayed on the screen, then the information is too small to read. Interactive infographics, on the other hand, can be designed to be the right size and legible on the screen, as well as neat by having various levels of information where more detailed information is available, as already mentioned.

Interactive infographics are therefore the most user-focused type of infographic of the three (static, interactive and motion) and have the highest ability to empower the user to be the ruler of their own information journey.

TYPE OF INTERACTIVITY

Interactive infographics can be distinguished in four types, based on Amit-Danhi and Shifman's (2020) distinction of interface types and their hypothesis that incorporating interactive infographic interfaces leads to an increase in user engagement with digital infographics. The four types include:

1. **Viewable** interactive infographics where interaction and performance are restricted to passively viewing the infographic and respective interactivity, i.e., communication only happens one-way, from the infographic to the user.
2. **Actionable** interactive infographics, which includes motion and users can choose when to begin and when to end the performance.
3. **Configurable** interactive infographics, which allows the user to change the information displayed but not the information itself; or change the data displayed but not the dataset.
4. **Constructible** interactive infographics, where users can change both the information displayed and the information itself; or the data display and the database itself.

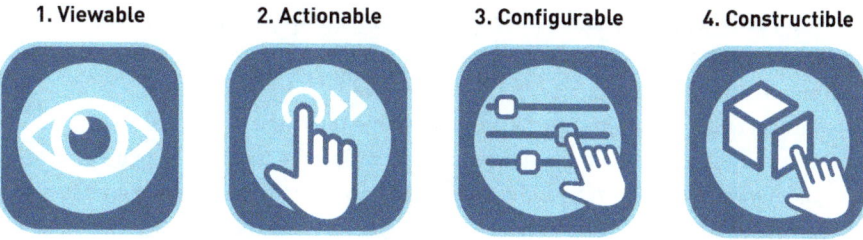

Figure 54 Types of interactivity

Interactive infographics can also be defined by the degree of interactivity, which ranges from low, to medium, to high interactivity, as defined by Langer and Zeiller (2017):

- **Low interactivity** refers to the manipulation of the interactive infographic without changing the infographic itself. Examples include zooming, mouseover effects for showing details, Next or Start buttons, etc.
- **Medium interactivity** refers to the manipulation of the interactive infographic by making changes and comparing information. Examples include using a timeline slider or menu items.
- **High interactivity** refers to the full exploration of the infographics and changing its content by inputting data, retrieving data, or filtering.

Interactivity, on the other hand, can be distinguished between interactivity with information, further divided in interactivity with fixed information and interactivity with dynamic information, and interactivity with the content (in agreement with Ali, 2021):

- **Interactivity with fixed information** relates to when the user interacts with fixed information simply by clicking or hovering on icons to explore more information and data. This type of interactivity is a good solution when the objective is that the information is viewed in a consecutive or chronological sequence.
- **Interactivity with dynamic information** relates to when the user views information that changes with time when clicking or hovering on a specific element in the infographic.
- **Interactivity with content** relates to when the user taps or hovers on a specific element (or icon) and navigates, for example, to another page to see the information about the topic they tapped on. It also relates to when users scroll, and as they scroll they see new and maybe unexpected information in a sequential and consecutive order, which is likely to grab users' attention for longer.

NAVIGATION AND INTENT

When developing interactive infographics, it is vital to consider the way users can interact with the infographics. Langer and Zeiller (2017) and Zwinger and Zeiller (2017) have identified three courses of action a user can take.

LINEAR INTERACTIVITY – This is where the users are restricted and can only follow a predetermined path, i.e., users do not have to explore the interactive infographic and visualisation by themselves. Users move forwards or backwards through a linear sequence and can only use navigation tools like Start, Stop, Forward, Backward, or Next. The creator/designer of the interactive infographics predetermines the step-by-step course, i.e., it is a creator-driven type of interaction.

NONLINEAR INTERACTIVITY – This is where users are not given a strict and prescribed ordering and are therefore required to be highly interactive with the infographic. Navigation tools for nonlinear infographics include: input box, data query, filter, or brushing. This means that it is a user-driven type of interaction. In this type of interactivity, the user has various different ways to explore the infographic. The exploration is also free, i.e., it does not have predefined navigation paths.

LINEAR–NONLINEAR INTERACTIVITY – As the name indicates, it is a combination of the linear and nonlinear approach, and consequently a mix between creator-driven and user-driven type of interaction. This means that the creator communicates the message using a predefined path, but the user also has a certain amount of selection. Navigation tools include interactive timelines, time controller and integrated navigation menu.

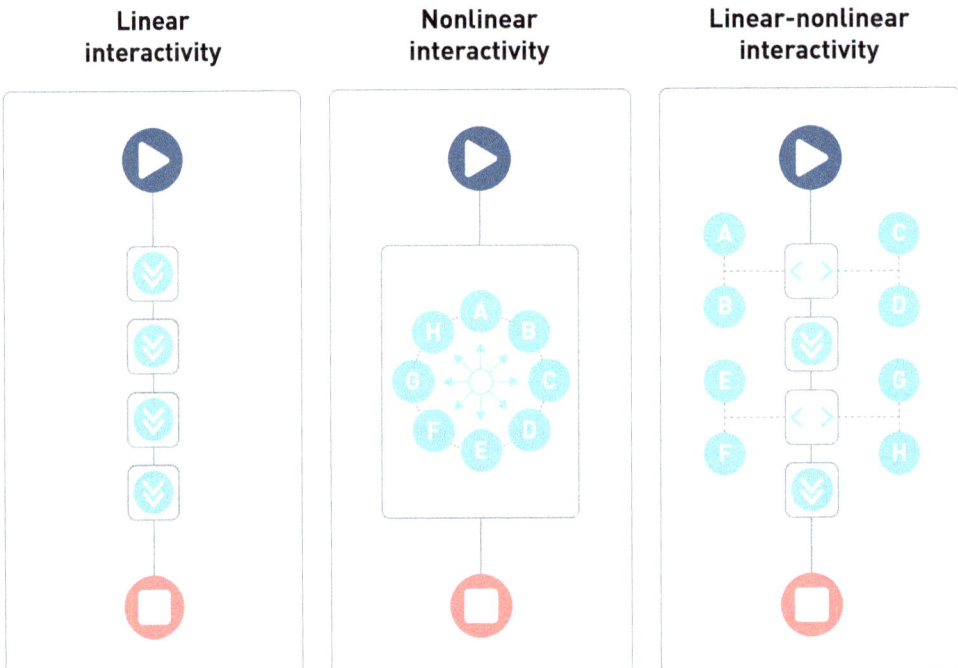

Figure 55 Ways users can interact with interactive infographics

Abdul-Majeed and Zeidan (2019) emphasise the fact that research studies often overlook an important design aspect of an interactive infographic, i.e., navigation through the infographic and what they define as 'navigability'. Navigability refers to the importance of navigations acting as a tool through which connections are made between gaps in the information provided and the extent to which users can know their current status and how they can move to other areas. Navigation levels are divided into two basic levels. The **horizontal level** is where users explore the interactive infographic in a sequential order without additional levels of information, i.e., users can navigate between each component of the interactive infographic and do not have to move into additional levels of each component in an infographic. The **in-depth navigation level**, in contrast, allows users to discover and interact with more infographic information than in the horizontal level, and users can also explore the relationship between any level.

When it comes to communicative intent, Langer, Zeiller and Zwinger identify four different formats:

- **Narrative intent**, where infographics are used to tell a story from a distinct point of view (e.g., anecdotes, personal stories, business stories, case studies).
- **Instructive intent**, where infographics are used to give step-by-step instructions to explain how things work, how to do something, or how events occur.
- **Explorative intent**, where infographics are used to allow users to discover the intention of the infographic by exploring actively and making sense of the information.
- **Simulative intent**, where infographic allow the readers to actually experience the intention of the infographic in a real-world situation.

INTERACTIVE ELEMENTS AND TECHNIQUES

There are several methods of interaction that allow users to control and/or manipulate information in interactive infographics. Options for control include, as listed by several researchers and authors, and of common knowledge to all of us: Start or Stop button; Forward or backward button; Menu item to select; Timeline or time controller; Filter, data request or input box.

Figueiras (2015) goes further by identifying interaction techniques. She conducted thorough research to build a comprehensive list of interaction techniques by evaluating 232 visualisations that were popular online and by studying the types of interaction used. From that study, 11 categories were identified: filtering, selecting, abstract/elaborate, overview and explore, connect/relate, history, extraction of features, reconfigure, encode, participation/collaboration, and gamification. Let us discuss each one separately, as this knowledge and understanding are needed before developing interactive infographics.

FILTERING OUT (i.e., eliminating or de-emphasising) items of no relevance to users, either by determining a range or a condition, is the most straightforward and quickest way to reduce complexity in interactive infographics. One of the biggest advantages of filtering is that the core information remains unchanged and can be displayed again as users choose. This ability to selectively hide and reveal pieces of information contribute to a reduction in cognitive overload and increase in focus and assimilation of information.

Figure 56   Interactive elements and techniques – Filtering out

SELECTING has a similar positive effect on cognition as filtering out. Instead of hiding information, users select the information of most interest to them and that they would like to emphasise in relation to other information that might still be of interest, but of less relevance. When the selection of information is combined with other interaction techniques, it maximises users' exploration of the information communicated to them.

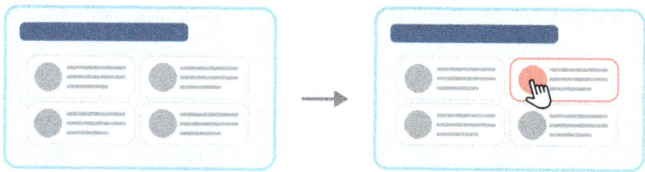

Figure 57   Interactive elements and techniques – Selecting

ABSTRACT/ELABORATE relates to the ability of users to regulate the number of stimuli that the interactive infographic provides by changing the amount of information that is displayed or emphasised. Within abstract/elaborate, **zooming** is distinguished as another interaction technique that, like filtering, helps reduce the complexity of information. For example, zooming-in enables users to remove irrelevant information or information of no interest from their visual field; allows users to see a smaller and more detailed view of the information without changing the infographic; and can even allow users to organise information into meaningful patterns. Zooming-out, on the other hand, enables users to have an overview of the information, especially after zooming-in, in order to go back to the information that was initially presented and that they might no longer recall. It is important to be aware, however, that zooming can lead to users losing their sense of position and context. It is therefore important to have a smooth transition between levels of zooming.

Another abstract/elaborate interaction technique is **details-on-demand**. Here users get additional information and detail when they select an item or component in an interactive infographic. This additional information is given on a point-by-point basis, which means that the view is not changed. A good example is the pop-ups obtained through hovering or clicking on the information, which allows users to access detailed information about a specific element. The details-on-demand technique is very common in narrative interactive infographics and can be either textual or graphic. By providing backstories and more detail, this technique is useful for increasing the level of user engagement.

Linking also falls under the umbrella of abstract/elaborate interaction techniques. It is often used to give access to external information and to a different set of information through, for example, hyperlinks that the user directly clicks on.

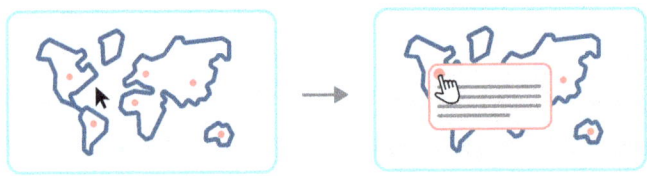

Figure 58   Interactive elements and techniques – Abstract/Elaborate

OVERVIEW AND EXPLORE refer to the entire set of information presented to the user at one time. Overview, as already discussed, is necessary to enable the user to refer back to the whole set of information when needed for checking, but also to identify patterns and themes. It is therefore important that users can overview the information right at the start. Understandably, in situations where presenting the entire set of information might be too overwhelming for users or there are limitations in terms of screen size, designers might choose only to display a limited amount of information at one time. This is acceptable as long as the entire information set is available should users choose to access it to examine and compare different subsets of information.

Figure 59   Interactive elements and techniques – Overview and explore

CONNECT/RELATE is an interaction technique that allows users to see relationships between pieces of information. For example, links between items can be highlighted by clicking on specific items, even if other techniques, such as colour coding, are already being implemented but are not enough on their own due to too much noise around the information of interest.

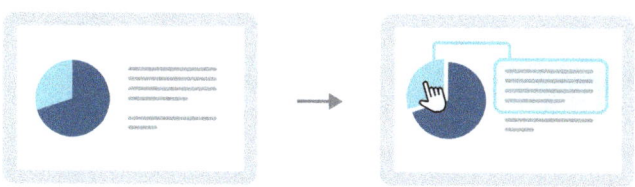

Figure 60   Interactive elements and techniques – Connect/Relate

HISTORY is an interaction technique of great importance that if dismissed will lead to user frustration but is often neglected by those developing interactive infographics. When the exploration of information is intrinsic to an interactive infographic, it will mean that users will follow a process with various steps and ramifications. It is therefore important to keep the history of actions to allow users to retrace their steps, as well as provide ways for users to recover from mistakes and/or refine their course of action.

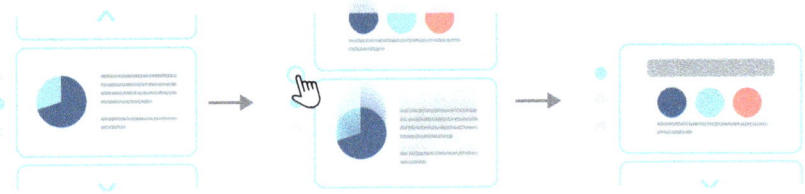

Figure 61   Interactive elements and techniques – History

EXTRACTION OF FEATURES is also important during such a complex explorative process. It is of great benefit to enable users to extract important information that they wish to share, keep a record of and/or analyse further. If information can be extracted, then the number of repeat actions within an interactive infographic can also be reduced.

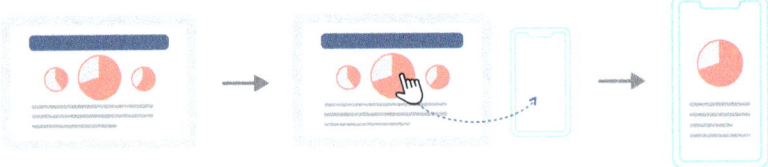

Figure 62   Interactive elements and techniques – Extraction of features

RECONFIGURE is an interaction technique that allows users to rearrange information as best helps them to process and understand it, or as they prefer to access it. This is done by changing the spatial arrangement of the information, when users are allowed to rearrange, for example, columns in a table, change attributes in a graph, and so on.

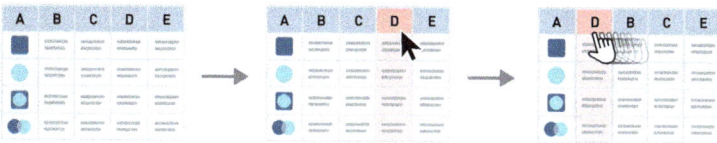

Figure 63   Interactive elements and techniques – Reconfigure

ENCODING technique allows users to change how information is represented in order to facilitate their understanding of the relationships within an information set or dataset. Changes can be made using different colours, different sizes and different shapes.

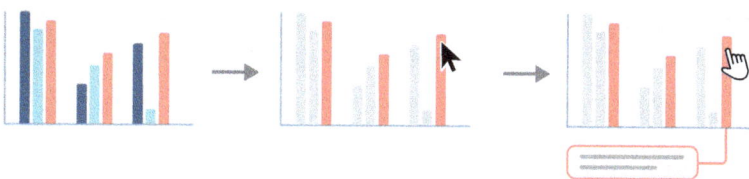

Figure 64   Interactive elements and techniques – Encoding

PARTICIPATION/COLLABORATION interaction technique empowers users to participate as a user of information but also as a contributor to the information they access. It also relies on the fact that there is more than one user contributing to the interpretation of the infographic, by sharing their insight. Usually, this collaboration is of a social nature and strategies are used that allow for social insights, such as tags, links, bookmarks, doubly linked discussions, graphical annotations, the traditional comments, and so on.

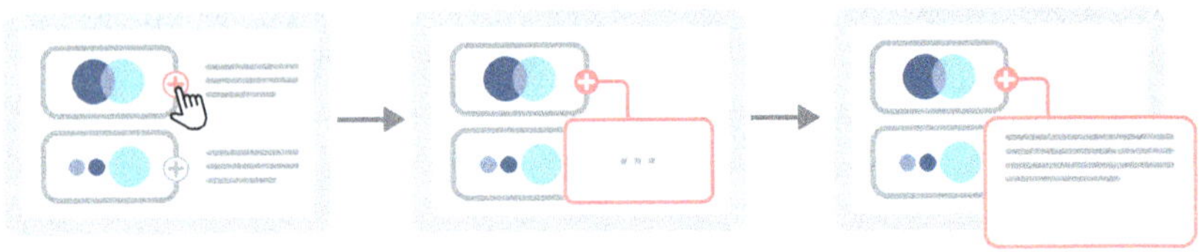

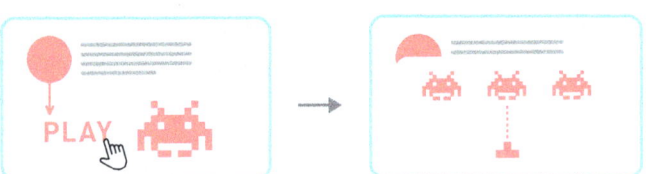

Figure 65 Interactive elements and techniques – Participation/Collaboration

GAMIFICATION is the most complex and least common interaction technique of all the 11 techniques discussed here. The main reason is that it is very time consuming, when less complex interactions are already time consuming themselves. This technique uses video game elements in non-gaming platforms to improve user experience and engagement. Therefore, an interactive infographic can only be considered to be gamification if game mechanics or game design patterns are included in addition to the other techniques discussed previously.

Figure 66 Interactive elements and techniques – Gamification

QUALITY DIMENSIONS

A number of complementary **quality dimensions** in user experience with interactive infographics in performing daily tasks have been identified by researchers such as Locoro and colleagues (2017). Quality dimensions include: information quality, interaction quality and design quality.

INFORMATION QUALITY further includes the formal qualities of intuitivity, elegance and attractiveness, and the substantial qualities of sinteticity, clarity and informativity. **Intuitivity** relates to the way information is organised in order to be immediately accessible to users (ideally at a glance). This immediacy can be enhanced if the information is familiar to the user, which in turn helps the user interpret the information better as well as make decisions with more certainty. Colour association and universal or cultural icons are good visual elements that can be used to increase familiarity. **Elegance** relates to the way visual information is presented and perceived by users, mainly the way information is designed in an orderly and clear manner through consistent and well-structured layouts, a clean and clear design, good use of white space and contrast between elements. **Attractiveness** relates to the aesthetic approach of visualising information that is capable of attracting and engaging users. In interactive infographics (and information visualisation in general) attractiveness is important to grab users' attention to the information in a first stage of information processing, at which point attractiveness does not depend on efficiency and effectiveness in interaction and use. **Sinteticity** relates to the capability of communicating information with the minimum number of elements possible by eliminating redundant information elements and avoiding density of information elements. **Clarity** relates to the ease of understanding and processing information. **Informativity** relates to the capability of communicating all relevant information and giving users a sense of completeness once they have accessed the information.

INTERACTION QUALITY includes usability and ease of use. **Usability** relates to the extent to which information can be used by users to achieve specific goals in an effective, efficient and satisfactory manner. **Ease of Use** refers to the user experience and how information helps users complete a task and make a decision.

DESIGN QUALITY relates to the value of the interactive infographic at the time of designing it (creator-focused), while the dimensions of information and interaction quality relate to the value of the interactive infographic at the time of using it (user-focused). Towards this end, instead of continuing with the contributions from Loroco et al. (2017), I am going to combine the work of several other authors to determine a set of design guidelines for the development of interactive infographics (e.g., Alshehri and Ebaid, 2016; Abdul-Majeed and Zeidan, 2019; Ali, 2021).

In order for interactive infographics to fulfil their primary goal of delivering content in a simplified and effective manner, several design guidelines should be considered. Some are of common application to all information design and visualisation. Some are more specific to interactive infographics due to functions beyond static communication of information:

1. A theme should be selected and made visible at a glance.
2. A good visual structure and hierarchy is imperative for users to grasp the information quickly.
3. Information should be organised in a logical order and with visual aids to help track information and understand relationships.
4. Sections should be used to chunk information and divide content to help information processing at the working memory stage.
5. Negative/white space should be used generously and with purpose to avoid information overload and allow the eyes to rest.
6. Simple and legible typography that contrasts well with the background is imperative. Sans serif fonts are favoured as these tend to be more legible on screen than serif fonts.
7. Important facts and areas should be distinguished visually and with colour, as appropriate.
8. Colour coding can be used effectively and sparingly to highlight key elements and/or make relationships.
9. Clear and familiar icons and symbols should be used when communicating data to help decomplexify the information.
10. Buttons and icons should be designed to make the interactive features clear, intuitive, predictable and easy to use.
11. Simple interactive infographics should be favoured to ensure accessibility, navigation and sustain engagement.
12. Consistency should be maintained between the design of the main information and its sub-levels (e.g., pop-ups).

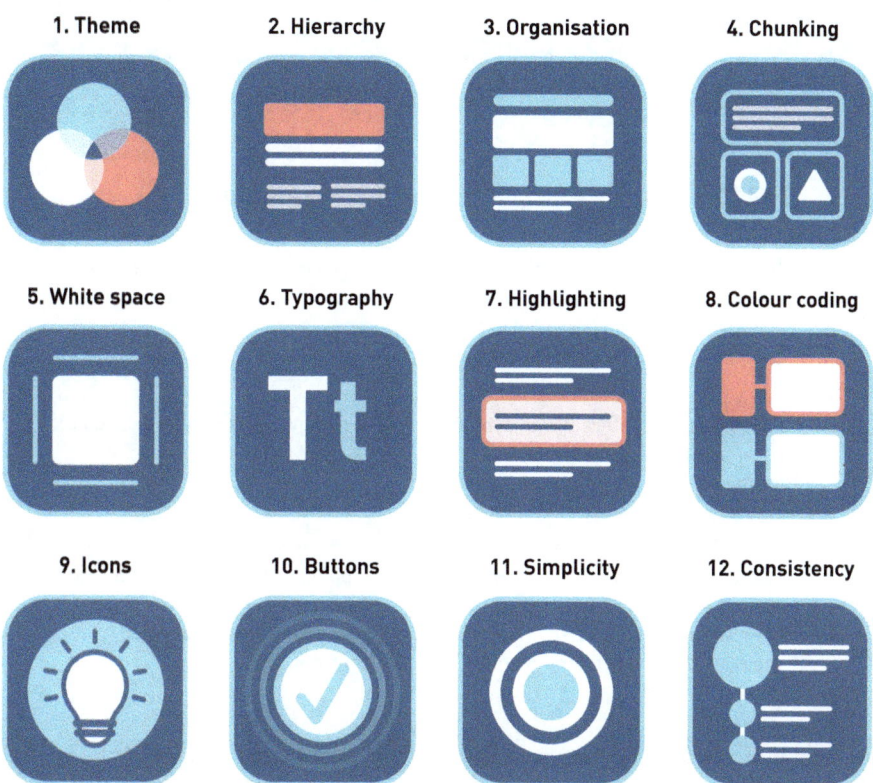

Figure 67 Quality dimensions

INTERACTIVE INFOGRAPHIC DESIGN

Top 15 principles

	Principle	Guidance	Example
1	Interactivity with fixed information	Enable users to interact with fixed information when the objective is that the information is viewed in a consecutive or chronological sequence.	Clicking or hovering on icons to explore more information and data.
2	Interactivity with dynamic information	Enable users to interact with dynamic information when the objective is for the user to view information that changes with time.	Clicking or hovering on a specific element in the infographic.
3	Interactivity with content	Enable users to interact with content when the objective is for the user to see more and unexpected information about a topic in a sequential and consecutive order.	Tapping or hovering on a specific icon, or scrolling down a page, to find new and unexpected information.
4	Linear interactivity	Restrict users to follow a strict path through a predetermined linear sequence if the content should be creator-driven and users are not meant to explore the interactive infographic and visualisation by themselves.	Navigation tools include Start, Stop, Forward, Backward, or Next.
5	Nonlinear interactivity	Enable users to be highly interactive with the infographic if the content should be user-driven, by giving users various different ways to explore the infographic freely.	Navigation tools include input box, data query, filter, or brushing.
6	Linear-nonlinear interactivity	Use a mix between creator-driven and user-driven type of interaction if the objective is to have a predefined path, but to also allow the user a certain amount of selection.	Navigation tools include interactive timelines, time controller, and integrated navigation menu.

Principle		Guidance	Example
7	**Filtering out and/or select**	Eliminate or deemphasise items of no relevance to users to reduce complexity.	By determining a range or a condition and/or by enabling users to select the information of most interest to them.
8	**Abstract/ elaborate**	Enable users to regulate the number of stimuli that is provided by changing the amount of information that is displayed or emphasised.	Zooming, details-on-demand and linking.
9	**Overview and explore**	Enable users to overview the information: right at the start; to refer back to the whole set of information to check; to identify patterns and theme.	Full set of information or limited amount of information at one time (with entire information set being available should users want).
10	**Connect/relate**	Allow users to see relationships between pieces of information.	Links can be highlighted by clicking on specific items.
11	**History**	Allow users to retrace their steps, recover from mistakes and/or refine their course of action.	Keeping the history of actions.
12	**Extraction of features**	Enable users to extract important information to reduce the number of repeat actions.	Functions such as share, keep a record of, and/or analyse further.
13	**Reconfigure**	Allow users to change the spatial arrangement of the information as best helps them to process and understand it, or as they prefer to access it.	Allowing users to rearrange columns in a table, change attributes in a graph, etc.
14	**Reconfigure**	Allow users to change how information is represented in order to facilitate their understanding of the relationships within an information set or data set.	Using different colours, different sizes, and different shapes.
15	**Information, interaction and design quality**	Effective infographics should be intuitive, elegant, attractive, synthetic, clear, informative, usable, and ease to use.	Clear structure and link between elements; sections to divide content; generous white space; legible typefaces; effective colour coding; clear icons/buttons, etc.

Figure 68   Top 15 principles for interactive infographic design

3.2 | MOTION GRAPHICS

Benefits of motion graphics

Techniques for motion graphic design

Motion graphic design. Top 15 principles

BENEFITS OF MOTION GRAPHICS

In addition to visualised information, motion graphics help to mentally visualise a process or procedure. This reduces cognitive load further, when we compare it to static information visualisation, where a process or procedure needs to be reconstructed from a series of static images. Motion graphics are particularly beneficial for instruction where step-by-step information needs to be followed.

Moreover, in static information visualisation, abstract elements (e.g., arrows) need to be interpreted and integrated with the visual information. In this case, there is always a risk of misinterpretation, while in a motion graphics the links between elements are clearer and information is fed in stages, which in turn also reduces cognitive load. Thus, we do not see all information at the same time, but over time, i.e., it is a temporal change that allows users to observe the development and progression of a process or procedure.

There is a disadvantage, though. As discussed earlier, during the working memory stage, our brain can only process a certain amount of information at one time. During a motion graphics, users will receive a large amount of information from the start to the end of the motion graphics, which is more than what our working memory is able to handle. Users' cognitive system is simply unable to process all the information presented in a motion graphics effectively. A way of tackling this could be as follows. When discussing interactive infographics, interactivity was highlighted as being able to help users understand and assimilate content better. While with motion graphics users are passive, receiving information as is the case with static infographics, enabling functions such as stopping, starting and replaying a motion graphics allow users to review and focus on specific parts and actions within a process or procedure. In addition, interactivity can be increased by allowing users to control speed, zooming-in and zooming-out, creating close-ups and alternative perspectives. Therefore, motion graphics that allow a higher degree of interactivity can facilitate perception and comprehension further.

Another angle to consider is the fact that if users are not fully engaged throughout the entire motion graphics, then the information that is communicated is not given the necessary active processing and can be missed. It is therefore important to keep users captivated by maintaining ongoing movement and using animation techniques that create a constant dynamic composition on screen without overwhelming users. Slowing time is also a good way of emphasising the importance of a certain piece of information, and accelerating time is a good way of speeding up a process that does not need detailed analysis and otherwise would take too long.

Using visual elements to direct users' attention to significant information is also important to captivate users, in addition to reducing cognitive load. This brings us to the importance of principles of visual perception. Our human visual system is very good at detecting motion, but it might struggle to perceive relationships in a moving display. This is even more problematic when two dynamic processes/ procedures are presented simultaneously but in different parts of the screen, as the eyes cannot follow both at the same time. Moreover, and taking into consideration Gestalt theory, elements in a motion graphics depend on where elements are positioned and whether and how these elements are moving as well. We need to use motion smartly to make sure that we attract users' attention to the important parts of the information we want to communicate, and not to where it is easy to perceive information in a motion graphics.

TECHNIQUES FOR MOTION GRAPHIC DESIGN

IMPLEMENTATION TECHNIQUES

In addition to the principles of animation, other features should be considered when designing a motion graphics. Taking in consideration Finke et al.'s (2012) five categories for the development of motion graphics, we should also consider: narration, animation, voice over, sound and focusing the viewer's attention. **Narration** in motion graphics is the spatial and temporal organisation of a narrated content in a linear sequence. Here the technique is to make sure that there is continuity and a relationship between the different scenes to guide users and hold their attention. For example, making switches between scenes smooth instead of quick and sudden, as the latter will only act as noise and disrupt users' attention. This relates to both the rhythm and length of the individual scenes and the pace between scenes.

Animation is then used to create a continuous change of visuals over time, i.e., in a temporal sequence, both in terms of content and/or space. Animation can also be used to make connections between content. To this end, we can use colour, for example, to show importance, hierarchy, dominance and meaning of the visual elements on the screen; to show contrast between visual elements and the background; and to connect and group visual elements together. We can also use form (shapes, lines, type, etc.), for example, to show quantities, direction, connection between elements and groups of elements, size and weight differences, rhythm, and so on. From my own research studies, I found that colour is the element that takes longer to get right. We need to go into the minute detail of what particular colour combinations, tones and contrast work on screen that do not add to **cognitive load** or **perceptual load**. That is, it is as difficult to process various colours at the same time (cognitive load) as it is to perceive colour combinations that do not have the right contrast or merge well together, and users cannot look or focus on that piece of information for any length of time (perceptual load).

What motion graphics offer that no other type of information graphics offer is voice over and sound. **Voice over** acts as a second channel of communication and perception, in addition to vision. Beyond helping to communicate information more clearly, a voice over can bring a more human and engaging side to the narrative. It is therefore important to get the voice over right and this is something that we can also spend a lot of time on just to get the right tone and pace. In addition, the voice over must link perfectly with the content on screen, i.e., what is being shown needs to match exactly with what is being said. Otherwise, this disconnection only serves to confuse users and increase cognitive load. If the tempo and match between visuals and voice over are not in syntony, then it can cause more harm than good. Arrows and highlighting techniques can also be used to further establish the connection between visuals and voice over.

Sound is different from voice over and includes background music that can be used to set the scene and create a thematic feel to the motion graphics. Depending on what is being presented, a calmer or more upbeat background music can be included. Ideally, this background music should also match, where needed, with how the content is being presented on screen. Background music is what we would call acoustic sound. We then also have causal sounds, which are defined as having the ability of cause and effect. These are sounds that have a direct link to what is being shown on screen and are usually in addition to the background music. For example, we might want to highlight a particular item among a group of items as we talk about each one, and as we do, a sound might pop-up.

All these are important factors if we want to communicate effectively, but also to grab **users' attention** and keep them engaged during the entirety of the motion graphics. Other techniques can include screen angles and zooming-in or zooming-out.

MULTIMEDIA PRINCIPLES AND TECHNIQUES

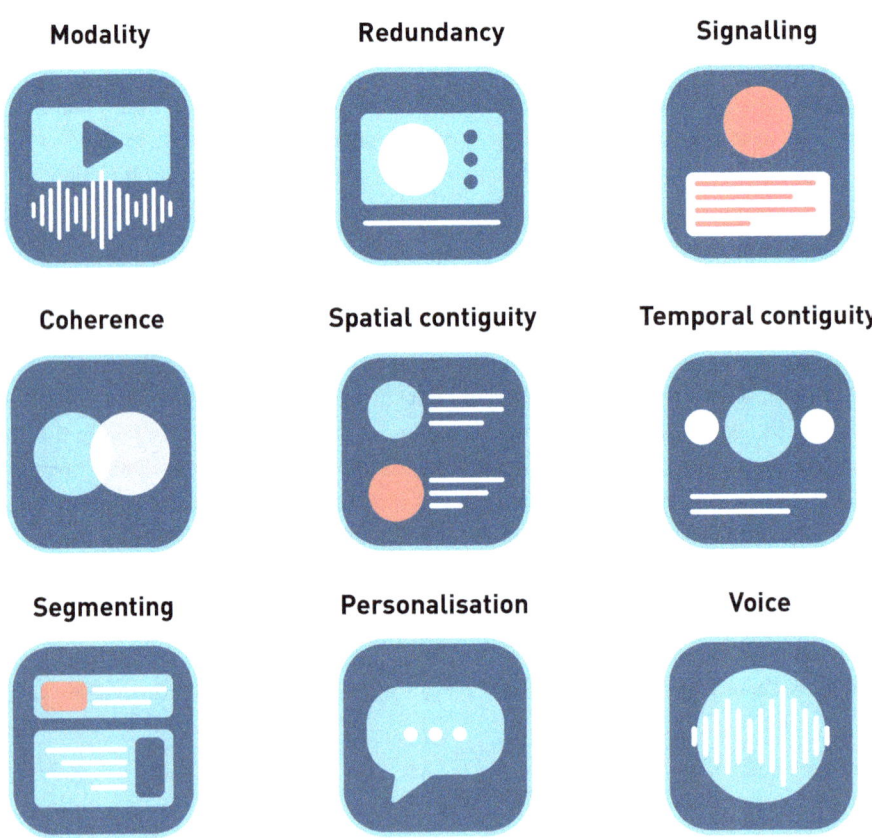

Figure 69   Multimedia principles and techniques

Other techniques can be considered, based on Mayer's (2014) multimedia principles, that can be adapted to motion graphics. Mayer's multimedia principles were developed to guide how best to structure multimedia learning experiences to enhance learner performance and comprehension. The ones discussed in this section are only those that can be adapted and guide the design of motion graphics.

The **modality principle** states that we learn better from a combination of visual and spoken information than visual and written information. This is not to say that written information should not be used on screen, but that too much text on screen in addition to visuals being presented in a temporal sequence will increase cognitive load and overwhelm users. Therefore, it is important to limit the amount of text on screen to key information that will add value to the content being communicated. For example, key words, numbers and percentages, steps, directions, etc.

The **redundancy principle** states that we should avoid presenting several sources of information simultaneously to users. Such redundant information will require users to coordinate all these sources at the same time, which will put a strain in their working memory and limit users' cognitive abilities. For example, having visuals, key text with visuals, voice over and subtitles, can be too much. One option is to have either visuals on screen or text on screen, but not both at the same time. Another option is to have subtitles and users can select to turn them on and off as they prefer. My own experience is that even if I am fluent in the language I am listening to, but subtitles are on screen as well, I am constantly drawn to the subtitles. This is very likely because they are signalling movement to which my eyes and brain are immediately attracted. It becomes very frustrating because during the time I am reading the subtitles without needing to, I am unable to focus on whatever else is happening on screen.

The **signalling principle** states that users become more effective when their attention is guided by cues that are added to relevant information or that highlight the organisation of the information. For example, highlighting important key words using type size, weight or colour; using lines, arrows and zoom to point out or focus on important information; or by having sections in a motion graphics to indicate that within the whole content there are subsections of information that should be noted.

The **coherence principle** states that users learn more deeply when extraneous and distracting information is excluded. Simplicity and minimalism are key here, where only information that is needed and that relates directly to the theme of the motion graphics should be included. Moreover, when choosing text, plain language and technical/key words that make sense and are easy to understand by the target audience should be used.

The **spatial contiguity principle** states that users learn better when corresponding words and visuals are presented on screen close to rather than far from each other, which directly links with the Gestalt principle of proximity.

The **temporal contiguity principle** states that users learn better when corresponding visuals and narration are presented on screen simultaneously rather than successively. Therefore, as already mentioned in relation to the voice over technique, voice over should be perfectly timed with the visuals as they appear on screen.

The **segmenting principle** states that users learn better when information is presented in segments rather than as a continuous unit, for example, by having information communicated in digestible, bite-sized chunks. Moreover, users also learn better if they can control the speed of the motion graphics, which in turn also increases interactivity, as previously discussed.

The **personalisation principle** states that users learn information better when using a voice over in a conversation style rather than a very formal style. For example, this can be achieved by using a more casual and positive tone of voice (and even an upbeat tone of voice in some situations) that is suitable for the target audience in question, rather than a tone of voice that is too formal, mono-toned and slow paced. It is also important to use plain language rather than complex or long words.

The **voice principle** is that users learn best from a human voice rather than a machine voice. To keep down costs and speed up the process, some motion graphics use a computer-generated voice. While some are quite high-quality, they still lack the warmth and human nature needed to engage users with the information provided. This is especially important in situations where user engagement is crucial, such as motion graphics for compliance with instructions, for behaviour change, to raise awareness, to aid decision making, and so on.

ANIMATION PRINCIPLES AND TECHNIQUES

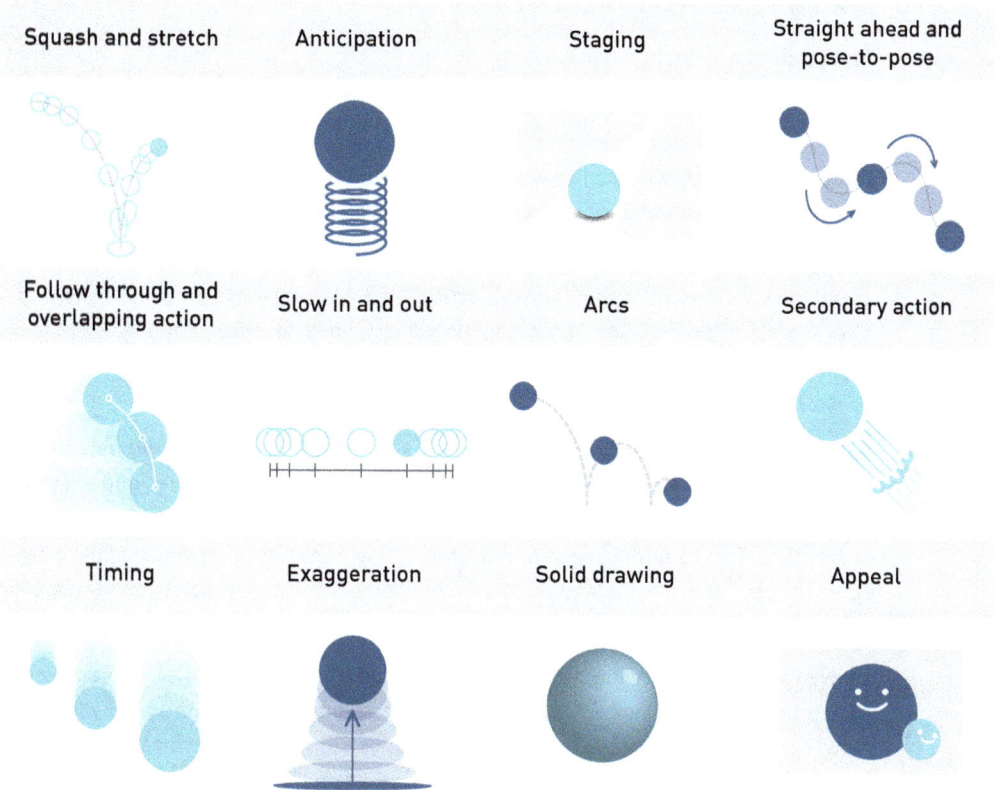

Figure 70   Animation principles and techniques

A similar adaptation from well-established principles in areas other than information design has been made between principles of animation and motion design. This is the case of the 12 Basic Principles of Animation introduced in the book *The Illusion of Life: Disney Animation* (Thomas and Johnston, 1995). There is a difference between human character animation and motion graphics. However, these 12 principles are applicable to motion graphics and have acted as the foundation of motion graphic design. The 12 principles are as follows.

Squash and stretch relates to how an object's weight, volume and flexibility is defined by distorting and changing its shape during an action. It gives a sense of speed and emphasises the object's path.

Anticipation relates to the preparation for an action, i.e., to anticipate the movement before it happens so that users are prepared for the action and the movement is also more believable. For example, moving back before moving forwards.

Staging is simply the arrangement or placement of the object on the screen to set the mood, the scene and focus, i.e., to direct users' attention to the important action.

Straight ahead and pose-to-pose relates to the two contrasting approaches to the creation of movement: straight ahead action is where each pose is created one after the other for fluid animation and movement; pose-to-pose action is where key frames are planned and then connected to each other, which leads to a more proportional animation.

Follow through and overlapping action refers to the fact that not everything moves at the same rate. Therefore, to give a real sense of flow, it is important that when a character stops, the hair, for example, continues to move for a little bit longer to create flow. To give a real sense of movement, it is important that when a character runs, the legs and arms, for example, move at different rates, which is an overlapping action/movement.

Slow in and out is quite important in motion graphics to give users a true sense of real-time movement. For example, in reality we cannot start at high speed without slowly gaining speed first, and we cannot abruptly stop without having a stage where we slow down first. Not creating a sense of slowing in and out will make the motion graphics look robotic, with sharp and abrupt movements (unless that is the intention at some points during the motion graphics).

Arcs relate to the visual path of action to denote natural movement. If we design objects to move in arcs rather than straight lines, it creates a sense of elegance, smoothness and rhythm.

Secondary action is needed to support primary action so that we create a good sense of reality and natural movement. For example, a person walking is a primary action, but if the person does not move their arms and other parts of the body (secondary actions), then the movement is not natural and looks robotic. It is important, however, that secondary actions add value instead of distracting users from the main actions and important information.

Timing is defined by the speed of an action through which we can define the characteristics of an object (e.g., when an object falls, the speed is different if the object is heavy or light), the personality of characters, and so on. Timing is what makes many of the other principles listed here work well.

Exaggeration relates to how we can accentuate the essence of an idea through the design and the action to make it more engaging and help communicate the information as clearly as possible. Exaggeration can be through the design as well as sound.

Solid drawing relates to adding a three-dimensional feel to an animated object to make it look more real. This can be achieved with 2D if perspective, shading and depth are added to the object to give the illusion of volume and depth.

Appeal relates to creating an object, character, scene or action that is captivating and the user enjoys watching.

All in all, considering the principles of information design and visualisation, together with the principles of motion graphic design, is the first step to assuring the appeal of a motion graphics. In the next section I list the top 15 principles for motion graphics.

MOTION GRAPHIC DESIGN

Top 15 principles

Principle	Guidance	Example
1 **Narration**	Make sure that there is continuity and a relationship between the different scenes to guide users and hold their attention (rhythm, length and pace of the individual scenes).	Smooth switches between scenes instead of quick and sudden.
2 **Animation**	Use animation to create a continuous change of visuals over time, i.e., in a temporal sequence, and to make connections between content.	Using colour and form to show hierarchy, contrast, quantities, direction, connection, size, weight differences, rhythm, etc.
3 **Voice over**	Make sure the voice over links perfectly with the content on screen.	Making tempo and match between visuals and voice over in syntony.
4 **Sound**	Use background music to set the scene and create a thematic feel to the motion graphics. Use sounds sparingly to emphasise content.	Calmer or more upbeat background music, depending on the theme. Sound effects to highlight an action.
5 **Modality**	Avoid too much text on screen in addition to visuals being presented in a temporal sequence.	Text limited to key information such as key words, numbers and percentages, steps, directions, etc.
6 **Redundancy**	Avoid presenting several sources of information simultaneously to users.	Having either visuals on screen or text on screen but not both; or having subtitles that can be turned on and off.
7 **Signaling**	Guide users' attention using cues that are added to relevant information or that highlight organisation of the information.	Highlighting important key words; using lines, arrows and zoom to point out or focus on important information; or having sections within a motion graphics.

Principle		Guidance	Example
8	**Coherence**	Remove extraneous and distracting information and favour simplicity and minimalism instead.	Reducing to information that is needed and that relates directly to the theme of the motion graphics.
9	**Spatial contiguity**	Present corresponding words and visuals close rather than far from each other.	Grouping elements according to Gestalt principles
10	**Temporal contiguity**	Allow users to change how information is represented in order to facilitate their understanding of the relationships within an information set or data set	Using different colours, different sizes, and different shapes
11	**Segmenting**	Present information in segments rather than as a continuous unit.	Using digestible, bite-size chunks; sections within a motion graphics; short motion graphics as part of a series.
12	**Personalisation**	Use voice over in conversation style rather than very formal style.	Using more casual and positive tone of voice; plain language instead of complex or long words.
13	**Voice**	Use voice over to bring a more human and engaging side to the narrative.	Using human voice instead of computer-generated voice.
14	**Control**	Enable users to stop and start the motion graphics and control speed to examine the specific information in depth.	Including speed control function.
15	**Speed**	Slow time to emphasise specific information or accelerate time to speed up processes or procedures.	Increasing or decreasing number of frames between actions.

Figure 71   Top 15 principles for motion graphic design

3.3 | INTERMOTION GRAPHICS

BENEFITS OF INTERMOTION GRAPHICS

Accessing and navigating complex information is a real cognitive challenge for users. As already discussed in this chapter, in addition to information visualisation and static infographics, access of complex information can be facilitated by using interaction or motion for information communicated on screen. Furthermore, design as a field is as fast paced as technology. This means that the foundation principles of motion graphics design have since expanded to, for example, principles of UX in motion. This is an excellent example of how motion and interaction come together, giving way to the fourth type of information visualisation output that I am going to call and define as **intermotion graphics** (i.e., animated graphic interactions for UI/ UX), where users can be both passive and active consumers of information.

For UI/UX services, motion alone cannot replace visible and interactive navigation, as happens with a motion graphics video, where the objective for communication of information is very different. However, motion can certainly be beneficial, in conjunction with interaction, to indicate to users the direction in which they are moving within a process, structure or hierarchy. Motion used to animate interactions can make information more intuitive and easier to understand than if no motion is applied.

But, as it is the case in every form of information communication, less is more when it comes to how much motion is used. Every time motion appears on the screen it attracts users' attention. Therefore, too much motion to animate interactions at the same time will lead to cognitive overload and information chaos. If more than one object moves at the same time on screen, it is difficult for users to track the movement. In cases where various objects and interactions need to be animated, then introducing them in sequence, for example, can avoid overwhelming users. In conclusion, if used well, motion can be the perfect solution to grab users' attention; if overused or misplaced, motion can distract, confuse and actually frustrate users.

In terms of accessibility, various considerations should be taken into account to make sure animated interactions are accessible. For example, we can include pause and play controls, provide options to reduce or increase motion, or even provide different versions of highly animated content where one will have a reduced level of animation.

With the need for interactive motion applied to user interfaces, contemporary designers have contributed to filling this gap in knowledge, although mainly informed by tacit knowledge. An excellent example is the work of Issara Willenskomer (2017) and his 12 principles of UX in motion. More general principles are given by Herman Morgan (2018) on the UX Matters website; by Page Laubheimer (2020) on the Nielsen Norman Group website; and by Nick Babich (2020) on the Adobe website. All these have been taken into consideration to help me define a set of principles for intermotion graphics presented later in this section.

PRINCIPLES OF INTERMOTION GRAPHICS

PRINCIPLES OF SPATIAL CHANGE

The **principle of transformation** relates to the process of a visual object changing its shape. This makes the object stand out and therefore grabs our attention. It also tells a story because it changes from one physical state to another. By guiding users through a series of separate key moments to get from the departure to their destination, transformation reduces cognitive overload, which in turn leads to more awareness, better processing of information and better retention. A good example is a submit button that changes into a progress bar as it is loading and then ends up as a check mark.

Figure 72 Principles of intermotion graphics – Transformation

The **principle of cloning** relates to the creation of new objects from existing ones during the time users are focusing on them. This also saves space and keeps the page uncluttered, which contributes to a reduction of cognitive overload. Here, too, users are guided from the departure to the destination. An example is when users click on a main button and that action generates other sub-buttons that emerge from the main button.

Figure 73   Principles of intermotion graphics – Cloning

The **principle of dimensionality** relates to the multiple interactive abilities animated objects can have, which are enabled or disenabled as needed. Users can click, drag, flip or fold an object, and all this is done in smooth transitions that are pleasant on the eye. Dimensionality also creates an illusion of depth, where objects exist spatially even if they are not visible (i.e., they are in front or behind each other in layers). Having this depth allows for a less cluttered page with multi-functions that users can decide to use or not, and this ability to choose prevents cognitive overload.

Figure 74   Principles of intermotion graphics – Dimensionality

The **principle of zooming,** as previously discussed, relates to when the object itself is not moving spatially, but simply scaling, i.e., enlarged or reduced. This allows for smooth transitions and brings up on the screen the relevant information as users need it, instead of having a page cluttered with all the information. Zooming can also help users understand the hierarchical information space: zooming-out shows fewer details and more objects, giving users the sense of being at the upper level of the hierarchy and seeing the whole structure; zooming-in shows more detail and fewer objects, creating the sense of going deeper into the hierarchy.

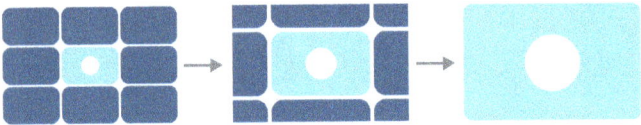

Figure 75   Principles of intermotion graphics – Zooming

PRINCIPLES OF VISUAL CHANGE

The **principle of visual cues** relates to animation being able to create visual cues that indicate to users where they are and what they need to do. Guiding users' attention to the right spot at the right time is especially important when there are many different functions in order to reduce cognitive effort and sense of loss. This is particularly relevant for novice users to whom the platform and display are new.

Figure 76   Principles of intermotion graphics – Visual cues

The **principle of visual feedback** relates to animation being able to tell users what is happening (i.e., current context and spatial location at a given time) through appropriate visual feedback, instead of users having to guess, which only mounts to frustration. Animated visual feedback therefore has the ability to make users feel that they are interacting with real information on the screen, and displays the result of this interaction. For example, when we press a button, we expect something to happen and our action to be confirmed, and using animation can do that easily, as well as make it more memorable. If something happens where that action leads to an unsuccessful outcome, animations are also ideal to provide information about that problem in a quick and clear way. Animated visual feedback is superior to static visual feedback because it grabs users' attention, whereas static visual feedback is ignored due to what is referred to as 'change blindness'. Animation can also be used in some cases for feedback even before users decide to commit to an action, for example, previewing the action by dragging elements to see what the result would be. All this supporting feedback reassures the users and decreases the time users would otherwise spend checking whether the action was successful or not, sometimes going as far as repeating the action for double confirmation.

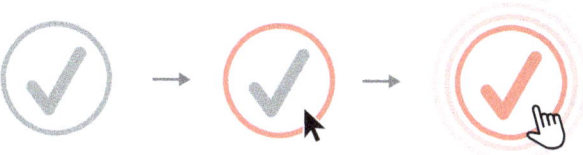

Figure 77   Principles of intermotion graphics – Visual feedback

The **principle of value change** relates to when numbers, for example, change to show progress, i.e., as the progress indicator bar loads. This is another type of visual feedback, but in this case, it is specific to informing users that their actions have had an effect, that the data is dynamic but something is still happening behind the scenes before the action is fully complete. So, the users understand that the data is dynamic. While this is happening in reality, it is labelled as a non-real-time event because the values change without user input to reflect a dynamic narrative. Good examples are progress loaders.

Figure 78   Principles of intermotion graphics – Value change

The **principle of neurofeedback** also relates to value change but occurs in real-time events where users are interacting directly with the objects to change values. An example is when users move a slider to set a price and the values keep changing as users keep moving the slider. Other examples include time display, scores in games, tracking one's fitness, to name just a few. In neurofeedback situations, users are empowered to become agents and decide the way they want to receive the information.

Figure 79   Principles of intermotion graphics – Neurofeedback

The **principle of mode change** relates to when animation is used to indicate that an object has changed to a different state. This is visual feedback that relates to the principle of transformation, but transformation of shape only and not in space and time, as discussed under principles of spatial change. For example, morphing an icon into another icon indicates more clearly that a transition has taken place instead of quickly and abruptly replacing one icon with another one. An example is a send icon that is morphed into a confirmation tick icon after it was clicked to signal the transition from Send to Sent.

Figure 80   Principles of intermotion graphics – Mode change

PRINCIPLES OF SPEED CHANGE

The **principle of timing** relates to the creation of realistic transitions. Independently of the animation style, there should be a balance in terms of speed when it comes to functional animation. Transitions should not be so fast that they feel abrupt and overwhelming to users, or so slow that they keep users waiting and irritate users. The balanced approach is to create animations that are slow enough that users notice the change, but quick enough that users remain engaged without the feeling that they are being delayed.

Figure 81   Principles of intermotion graphics – Timing

The **principle of easing** relates to the acknowledgement that all live objects do not move at a consistent speed and therefore we should follow the same rules for intermotion graphics, as previously discussed for motion graphics. By using easing (acceleration or deceleration), we can create natural movement and make objects look less robotic and artificial. To this end, it is also important to use nonlinear movement, i.e., animate objects move at different speeds. For example, gravity tells us that an object starts at a slower speed at the beginning of a fall and speeds up as it falls. This should apply to intermotion graphics with the intention that movement is seen by users as natural, invisible and non-distracting.

Figure 82   Principles of intermotion graphics – Easing

PRINCIPLES OF HIERARCHY AND STRUCTURE CHANGE

The **principle of offset and delay** relates to the relationship between elements and hierarchy. For example, objects might appear one by one to create the impression that they are part of a whole but with different roles and/or levels of importance, or to be seen following a specific order. Therefore, even before the user fully registers what the objects are, they are already being told part of the nature of the objects, which is then complemented by the visual design. For example, we can show a heading and image first to give part of the information, and only then a button that they need to click related to that heading and image (i.e., on delay). This will not only grab users' attention, but also make them notice the interactive button that will appear unexpectedly and therefore emphasise its importance at the time action needs to be taken.

Figure 83   Principles of intermotion graphics – Offset and delay

The **principle of parenting** also helps define relationships between objects by linking their properties and establishing a direct connection. In this case, I am referring to a parent object and a child object. For example, the parent and child objects might be connected by size, position, colour, shape, scale, value, rotation, opacity, etc. So, when users interact with the parent object, the child object should move accordingly. This is an excellent way of using motion to communicate to users how objects are linked and what the relationship between them is.

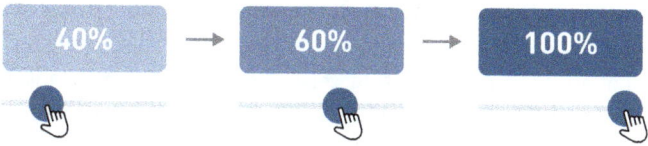

Figure 84   Principles of intermotion graphics – Parenting

The **principle of overlay and obscuration** relates to the ability to place one layer over another, as mentioned previously in relation to the principles of space change. This increases spatial hierarchy and unclutters the page. It also shows depth by using motion on objects that are dependent on location and exist behind or in front of others. It is more intuitive to users to hide, visually and cognitively, objects that are not needed at a given time. For example, the ability to archive information by storing it in an archive folder that can be retrieved if needed, is an example of overlay. Techniques for obscuration include blurring and overall object transparency (e.g., confirmation dialogs, navigation menus, etc.).

Figure 85   Principles of intermotion graphics – Overlay

The **principle of parallax** relates to objects moving at different speeds to create an illusion of depth, where objects moving slower are perceived as being further away, and objects that are further away are perceived as being less important. This means that they are perceptually and cognitively considered as secondary objects. On the other hand, objects that move faster call our attention and are perceived as more important. Visual elements like text, buttons and images are usually given higher priority and therefore move faster.

Figure 86   Principles of intermotion graphics – Parallax

INTERMOTION GRAPHIC DESIGN

Top 15 principles

	Principle	Guidance	Example
1	Transformation	Guide users through a series of separate key moments to get from the departure to the destination.	A submit button that changes into a progress bar as it is loading and then ends up as a check mark.
2	Cloning	Create of objects from existing ones during the time users are focusing on them to save space and unclutter the screen.	When users click on a main button and that action generates other sub-buttons that emerge from the main button.
3	Dimensionality	Use the multiple interactive abilities of an animated object to create an illusion of depth and keep the screen uncluttered.	Allowing users to click, drag, flip, fold an object and in a manner that all is done in smooth transitions.
4	Zooming	Use zooming to create smooth transitions and to bring up on the screen the relevant information as users need it.	Zooming out to show the whole structure; zooming in to allow users to go deeper into the hierarchy.
5	Visual cues	Create visual cues that indicate to users where they are and what they need to do.	Using colour, signs and visual elements to guide and direct.
6	Visual feedback	Use animation to tell users what is happening through appropriate visual feedback.	Using animation when users press a button to show that something is happening and to confirm the action.
7	Visual change	Use animation to show progress and that users' actions have had an effect.	Using animated progress load bars.
8	Neurofeedback	Use animation to change value in real time and as users interact directly with the objects.	Using slider to set a price, display time, scores in games, tracking one's fitness.

Principle		Guidance	Example
9	Mode change	Use animation to indicate that an object has changed to a different state.	Morphing an icon to another icon, such as a send icon that is morphed into a confirmation tick icon after it was clicked.
10	Timing	Create realistic transitions independently of the animation style, i.e., transitions should not be too fast that feel abrupt and overwhelm users, nor too slow that keep users waiting and irritate them.	Creating animations that are slow enough so that users notice the change, but quick enough that keeps users engaged.
11	Easing	Use acceleration or deceleration and non-linear movement to create natural movement and make objects look less robotic and artificial.	For an object that is dropping, starting it at a slower speed at the beginning of the fall and then increasing the speed as it falls.
12	Offset and delay	Show objects one by one to create the impression that they are part of a whole but with different roles and/ or levels of importance; or that these objects follow a specific order.	Showing a heading and image first to give part of the information, and only then a button that they need to click related to that heading and image (i.e., on delay).
13	Parenting	Link objects by their properties to establish a direct connection between a parent object and a child object and to also establish the relationship between them.	Connecting parent and child objects by size, position, colour, shape, scale, value, rotation, opacity, etc. and making them move together accordingly.
14	Overlay and obscuration	Place one layer over another to increase spatial hierarchy and show depth by using motion on objects that are dependent on location and exist behind or in front of others.	Archiving information that can be retrieved if needed. Blurring and overall object transparency.
15	Parallax	Move objects at different speeds to create an illusion of depth where objects moving slower are perceived as being further away and less important; and objects that move faster call our attention and are perceived as more important.	Moving important text, buttons and images faster to give them higher priority.

Figure 87   Top 15 principles for intermotion graphic design

OVERALL CONCLUSION

With the high increase of internet and digital users, even more so after the Covid-19 pandemic, so the amount of information and data that we are exposed to on a daily basis has also increased, to the point of being overwhelming. Communication that is interactive and/or in motion allows information to be communicated in a more digested form; allows information to be filtered in such a way that while it is in high volume, it is presented in layers and chunks for easy access and assimilation; and allows users to be active consumers of information and have agency to decide how much and how quickly they want to absorb the information that is available to them. In addition, this agency and ease of use also leads to more engaged users and gives them the power to understand information better, to learn better, to change behaviour, to make decisions, and so on.

But as we often say when referring to typography, i.e., good typography should be invisible and not call attention to itself, the same applies to interactive infographics, motion graphics and intermotion graphics. All forms of digital information visualisation should be highly effective but invisible to users. Users should not notice that they are looking at the interaction and/or motion, but instead notice the excellent experience that they are having. To be able to achieve that, designers and creators should bear in mind the techniques and top principles listed and explained in this chapter, to guide their design.

WANT TO LEARN MORE ABOUT THIS TOPIC?

Langer, Zeiller and Zwinger | Various papers

Papers published by these researchers are some of the very few discussing in detail interactive infographics. Their research and findings are essential to understand and design interactive infographics. See in particular, Langer and Zeiller (2017) and Zwinger and Zeiller (2017).

Finke, T., Manger, S. and Fichtel, S. 2012. *Informotion: Animated Infographics*. Berlin: Gestalten.

In my opinion, this is the only book that looks closely at the fundamentals of motion graphics from a design perspective, taking into consideration how viewers process information. The book is also very tastefully designed.

Mayer, R.E. 2014. *The Cambridge Handbook of Multimedia Learning* (2nd edn). Cambridge Handbooks in Psychology. New York: Cambridge University Press.

This book is of direct relevance to the design of motion graphics. It discusses in detail principles of effective multimedia instruction based on solid theory and research to explain why certain practices succeed and some fail.

Willenskomer, I. 2017. Creating usability with motion: the UX in Motion Manifesto. *UX Motion* [Online].

Willenskomer's Motion Manifesto is very well written and has very good real-life examples. It helps readers understand the relationship between interaction and motion when it comes to UX and its great benefits if well applied.

RELEVANT REFERENCES

INTERACTIVE INFOGRAPHICS

Authors that have informed this section, include: Dur, 2014; Figueiras, 2015; Alshehri and Ebaid, 2016; Langer and Zeiller, 2017; Locoro et al., 2017; Zwinger and Zeiller, 2017; Abdul-Majeed and Zeidan, 2019; Amit-Danhi and Shifman, 2020; Ali, 2021.

MOTION GRAPHICS

Authors that have informed this section, include: Thomas and Johnston, 1995; Babic et al., 2008; Finke et al., 2012; Mayer, 2014; Crook and Beare, 2015.

INTERMOTION GRAPHICS

Authors that have informed this section, include: Willenskomer, 2017; Morgan, 2018; Babich, 2020; Laubheimer, 2021; several authors on the Nielsen Norman Group website.

4

User-Centred Research
Methods for Visualisation

INTRODUCTION

CONTEXT

The word 'research' in a practice-led field such as design is often seen as scary by students and unnecessary by creatives, who see it as expensive and time consuming, and expertise that they do not have. However, design has and should have the users' needs and expectations at its core. Failing to acknowledge this is failing to 'design'.

I often use the expression 'design for the user and with the user' and this is why research is important. In information design, and information visualisation in particular, information outputs need to be accessible, clear, easy to use and engaging to the target audience they are intended for. Moreover, once implemented, they need to be robust and withstand exposure in real-life scenarios. This can only be achieved if research informs the creative design process at every stage.

This chapter focuses on the different stages of the combined research and design process, and suggests user-centred design research methods that can be used in each stage. It describes methods that can be flexible, depending on the time and budget available for the project in question, with the caveat that the more in-depth research we conduct the better, and the more reliable and the more applicable to practice the design output can be. And while in-depth research will require more time, it does not need to be expensive. Equally important is the fact that research is essential to avoid producing an output that does not meet user needs, is biased by reflecting the designer's personal insights, and can end up being more counterproductive than helpful.

All in all, we cannot design for users without understanding their needs, their expectations and the context in which they will be exposed to and use particular information. This can only be achieved by user-centred research, and not doing so often leads to poorly designed information, negative impact on user performance and user engagement with information. It is also actually more expensive and time consuming to design without research than if research is conducted.

In this book I define user-centred design for visualisation as a research and design process that: (a) is collaborative and iterative in nature (for the user and with the user); (b) focuses on the user needs and expectations at every stage of the process; and (c) uses a variety of research methods, as needed, for each particular user cohort and budget.

I also define the stages of the research and design process, within which I am going to discuss the various research methods, as being:

1. **Discovery** – the stage where we understand the user needs and expectations, the context in which they will access information, and identify any potential problems. This is the stage where we should look at previous research and completed studies (such as the ones presented in Chapter 5, for example). Problem discovery will avoid frustration later and mitigate financial risk and loss.

2. **Exploration** – this is the stage where more direct input from the user happens in terms of ideation and creative design solutions, i.e., users as co-creators of information design.

3. **Development** – this is the most design-focused stage, where a series of design development and iteration steps happen (as many as needed or as many as time and resources allow). The creative design is solely developed by the information designer, but users test the materials and give feedback which is then considered during the following design iteration(s).

4. **Evaluation** – although during the development stage design solutions are being evaluated, this is the final stage where we implement the design output, which should be the optimum design output (or as close as possible). We also test it in real-life contexts and with a sufficient number of users to be able to conduct statistical analysis that further helps validate output effectiveness. In this stage, although to a lesser extent, if any problems are still identified, there is a last opportunity to rectify them before the final launch of the design output.

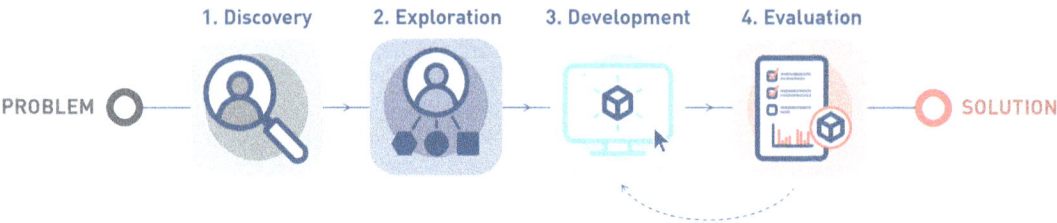

Figure 88   Research and design process

BASICS OF RESEARCH

Research can be generated in two formats: quantitative and qualitative research. **Quantitative** research produces findings and data with numerical values, logic and objective data. Numerical results can be calculated using statistics to define whether there is a statistical significance, which in turn strengthens the reliability and validity of the findings. Quantitative data can then be visualised (i.e., data visualisation), for which information visualisation principles are imperative, such as the ones presented in Chapter 2. Quantitative research answers questions such as 'how many?', 'how much?', 'how often?'. **Qualitative research**, on the other hand, explains user behaviour and why something happens. This means that results are not numerical, but instead include users' opinion, comments, feelings, preferences, and so on. The best research findings are, without doubt, the ones generated through combining both quantitative and qualitative research methods.

Another distinction is between primary and secondary research. There is always a lot of confusion regarding these because secondary research actually takes place before primary research. So, the denomination of 'primary' and secondary' is not associated with what takes place first and second. Let us therefore follow the research and design process and understand the sequence of steps. **Secondary research** relates to research that was conducted by someone else and that we should read and analyse before conducting and planning our own research. Secondary research is, in essence, the review of literature and other sources (published statistics, expert analysis and reports, documentaries, personal diaries, etc.) that we analyse to set the context of our own research, to identify what has been researched already and what still needs to be researched. Secondary research does not need to be

Quantitative vs Qualitative research

Differences	Quantitative	Qualitative
1 **Data definition**	Numerical data that can be quantified and measured.	Observational data that is descriptive and interpretative.
2 **Questions asked**	How much? How many? How often?	Why? How? What is the reason for something?
3 **Data type**	Fixed and universal, factual data.	Subjective and dynamic, open to different interpretations.
4 **Data examples**	Accuracy scores, time efficiency, age, height, error numbers.	Qualifications, religion, aesthetic ratings, user opinion.
5 **Data analysis**	Can be compared using statistics to determine significance.	Interpreted by the identification of themes and categorisation of data.
6 **Research examples**	A/B testing, surveys, experimental comparison.	Interviews, opinion questionnaires, focus groups, task analysis.
7 **Strengths**	Often easy to collect and analyse, and can be statistically compared.	In-depth data that explores bigger meanings and relationships.
8 **Weaknesses**	May not investigate the bigger picture and limits interpretation.	Harder to analyse and unable to determine statistic significance.

Figure 89   Quantitative vs qualitative research

exclusively focused on our design problem. Looking at research in other areas can bring more creativity, clarity and strategic problem solving to our design problem. I highly recommend designers/researchers to do this, especially as information design is still a growing field of research and needs to be informed by other relevant areas. **Primary research**, on the other hand, is the original research that we conduct ourselves to solve the design problem, identify user needs and find an optimum design solution. Primary research methods include all the methods discussed in this chapter.

Research can also be divided into generative and evaluative research. **Generative research** happens at the start of the research and design process. It is when researchers/designers identify and define a problem and then generate possible solutions. This means that in generative research we start (and should start) with various possible solutions based on secondary research, and then as the problem and user needs become clearer, we start using primary research methods to narrow down the range of solutions. **Evaluative research**, on the other hand, takes place towards the end of the research and design process. It is used to test, iterate and refine the optimum design solution. In sum, generative research corresponds to the 'Discovery' and 'Exploration' stages in my model. Evaluative research corresponds to the 'Development' and 'Evaluation' stages of my model, because evaluative research should happen at the early stages of design prototyping and not just at the end.

4.1 | DISCOVERY STAGE

1. Discovery

Online questionnaires

Interviews

PROBLEM

Focus groups

Diary studies

Visual survey and heuristic evaluation

Eye-tracking

Task analysis and observation

Figure 90 Discovery stage diagram

ONLINE QUESTIONNAIRES

Online questionnaires are used to collect information that is then used for analysis. Online questionnaires are not used to change user behaviour but to discover important information related to users by asking them directly about the points concerned with the design research. Online questionnaires are best used when the information required is brief and straightforward and does not need face-to-face interaction.

The success of a research questionnaire depends on three things:
(1) Response rate – how many questionnaires are returned; (2) Completion rate – how many are fully completed; and (3) Validity of responses – how honest and accurate are the answers. Independently of how well the questionnaire is designed, if they are not completed and submitted, they have no value.

Therefore, when devising a questionnaire, the following should be considered. What are the **capabilities of the respondents**? What is their literacy (i.e., their reading ability), proficiency with the language and their computer literacy to complete the questionnaire online? What is their sight capacity, i.e., blindness, vision disorders, age? Are they vulnerable, i.e., are they young people or children, the elderly, do they have learning disabilities? **Respondent motivation** also needs to be considered. Are they interested and enthusiastic about the topic? Do we need to encourage participation through a prize draw, for example? Is the way the online questionnaire is distributed convenient for them, and do they have the time to complete it, or is it too long? This leads to the **structure of the questionnaire**, where consideration should be given to both the number of questions asked and the amount of time needed to complete the questionnaire. Consideration should also be given to the ease of answering questions by making sure that questions are not ambiguous, and a suitable question format is used.

With this in mind, we should always assume that people will be reluctant to complete an online questionnaire. We must therefore think how we can entice them to make the effort to complete it. We should always pilot the questionnaire ourselves by putting ourselves in the respondents' shoes and reflect on why people should bother to complete it.

In addition, and considering that questionnaires are a piece of information design themselves, the following should be taken into account. **Appearance** is usually the first thing to which respondents react. Therefore, a clear and professional appearance will encourage participation and increase response rates. **Design features** can then help to improve questionnaire appearance. For example:

- Generous spacing to make the reading easier and allow respondents to discern between the different sections, related questions and answer spaces.
- Consistent positioning of response boxes to speed up completion and increase accuracy and consequently avoid missed responses or responses in the wrong place.
- Good choice of typefaces to maximise legibility.
- Good use of typographic features to differentiate instructions from questions (e.g., regular vs bold, lowercase vs capital, larger and smaller type size).
- If a long questionnaire is necessary, pages should be left unnumbered, and even more attention should be paid to layout and appearance.

The order in which questions are presented is also important to engage respondents and make sure the questionnaire is completed and submitted. The most crucial stage might be the introduction, where participants will be convinced (or not) to complete the questionnaire. Usually, questions on personal details appear first as these are answers that respondents will know without much effort. Starting with easy questions also provides the opportunity for practice in answering. But essential information should also appear early in case users lose focus later when completing the questionnaire, while unimportant questions are best left at the end.

Online questionnaires are often seen as 'second class' as a research method, which to me is an unfair assumption. If well-structured, well-designed, and developed with a clear purpose in mind, questionnaires have many advantages over other methods. They are economical, allowing a lot of data to be collected at a low cost. They are also relatively easy to arrange because respondents are free to choose when and where they complete the questionnaire. The delivery of the questionnaires is standardised, which minimises interpersonal factors affecting the way questions are asked. Data is accurate with online questionnaires because the answers can be directed into a data file, eliminating human error. Online questionnaires also facilitate accessibility, with people with sight impairment or learning difficulties being able to adapt the online questionnaire or form of completion as needed.

HOW TO CONDUCT QUESTIONNAIRES?

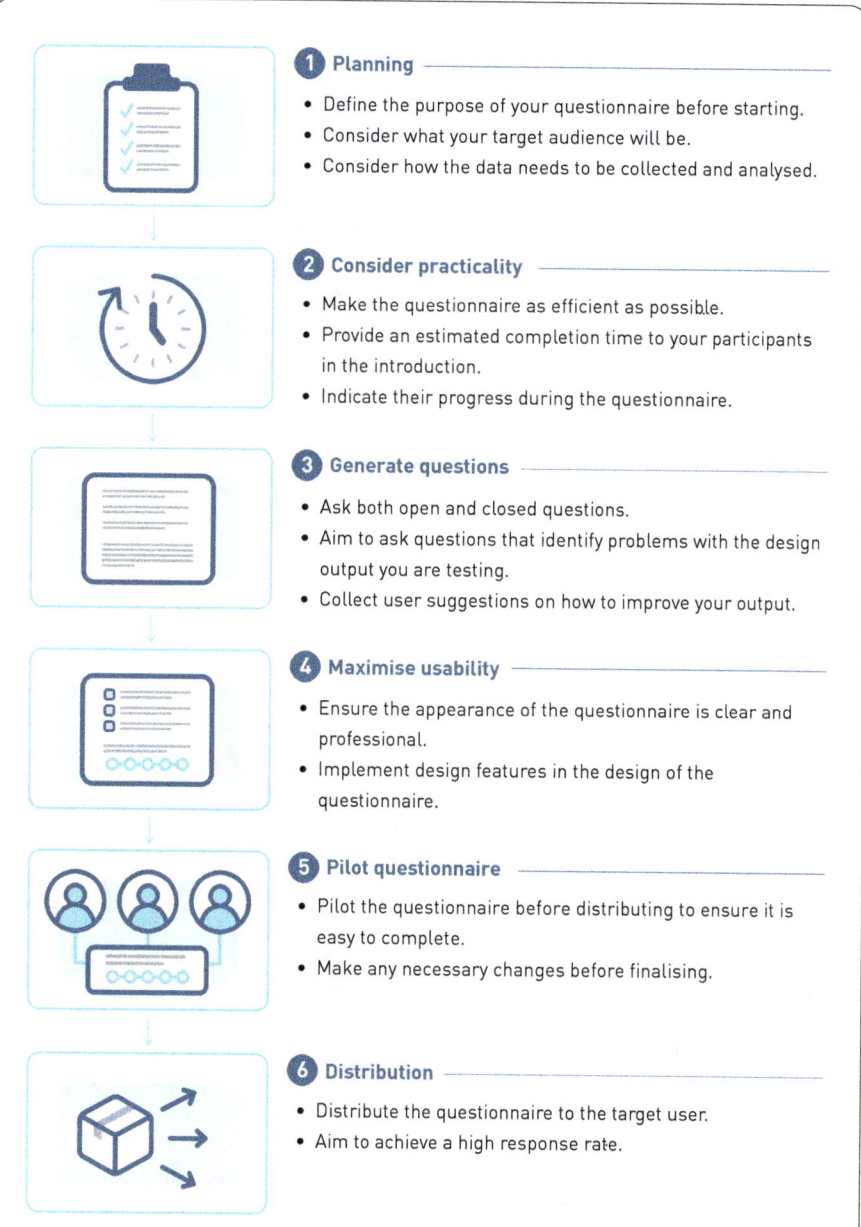

1 Planning

- Define the purpose of your questionnaire before starting.
- Consider what your target audience will be.
- Consider how the data needs to be collected and analysed.

2 Consider practicality

- Make the questionnaire as efficient as possible.
- Provide an estimated completion time to your participants in the introduction.
- Indicate their progress during the questionnaire.

3 Generate questions

- Ask both open and closed questions.
- Aim to ask questions that identify problems with the design output you are testing.
- Collect user suggestions on how to improve your output.

4 Maximise usability

- Ensure the appearance of the questionnaire is clear and professional.
- Implement design features in the design of the questionnaire.

5 Pilot questionnaire

- Pilot the questionnaire before distributing to ensure it is easy to complete.
- Make any necessary changes before finalising.

6 Distribution

- Distribute the questionnaire to the target user.
- Aim to achieve a high response rate.

Figure 91 Online questionnaire process

INTERVIEWS

Interviews should be used when the aim is to gather information in direct contact with the user and to collect first-hand personal accounts, for example, to understand in-depth opinions, feelings, emotions and experiences, and to understand how and how often information is accessed and what factors contribute to it. Interviews are also good to acquire privileged information from people who can give valuable insights.

It is pointless planning to conduct interviews unless there is access to the users and target audience. For example, do they have availability, do they need approval (e.g., children, vulnerable adults). Costs also need to be considered. For example, if interviews require considerable travel costs and are time consuming if in distant locations, can they be conducted online while still preserving the human contact? Interviews also involve transcription time and costs.

The sequence of questions in interviews can be classified as follows.

STRUCTURED INTERVIEWS: There is careful control in relation to the format of the questions and answers. It is more like a questionnaire, but one that is conducted in person, where a list of predetermined questions is asked, and respondents can only give limited-option and short-answer responses. Structured interviews are useful with large projects involving a lot of straightforward questions. Structured interviews also facilitate data collection.

SEMI-STRUCTURED INTERVIEWS: These still involve a clear list of questions to be answered but are more flexible in how the order of the topics is presented and discussed. Respondents can therefore express ideas and speak more widely as the answers can be open-ended and the emphasis is more on points of interest. With this structure, it is also possible for the interviewer to develop and change the interview questions from one interview to the next. That is, questions can change because of information gathered in the previous interviews.

UNSTRUCTURED INTERVIEWS: Also called in-depth interviews. The interviewer begins by asking a general question or gives a general topic, and then encourages respondents to talk freely. This means that respondents develop their own ideas, and the interviewer uses an unstructured format where the direction of the interview is determined by the previous answer given. To conduct this type of interview it is important to be equipped with probes such as: 'What do you mean specifically?', 'Would you like to tell me more about that particular point?', 'Is there anything else you want to add?'.

The success of an interview relies as much on the good structure and quality of the questions as on the skills of the interviewer and how the interview is conducted. It is important to be attentive and to look engaged in the conversation. Therefore, it is better to record the interview than to write notes as the interview is conducted. It is also important as an interviewer to be able to tolerate silences and not jump quickly into the next question because respondents might still have more to say. Equally, there might be times when the interviewer needs to prompt respondents to speak and expand on their answers by, for example, repeating the respondents' last few words to invite them to continue; repeating the question for further clarification or to allow more time to think about the response; offering some examples to help with answering the questions; or simply remaining silent to use a pause as a prompt to indicate that respondents have time to put their ideas together. Probes, on the other hand, can be used to politely ask respondents for examples, for more clarification and for more detail. Finally, it is very important not to be judgemental as an interviewer by not revealing personal values, disgust, surprise or acceptance through facial gestures. It is also important to assure respondents that they can choose not to discuss any sensitive issue to them.

HOW TO CONDUCT INTERVIEWS?

1 Recruitment

- Contact participants in advance to organise a time and location.
- Inform the participant how long the interview will take.
- Create a comfortable setting for the interview.

2 Introductions

- Allow the interviewer to introduce themselves and the aims of the research.
- Set a relaxed tone for the rest of the interview.

3 Starting the interview

- Start with an easy question related to the participants' relationship to the topic, or using stimulus material.
- You should aim to relax the participant.

4 Monitoring

- Identify the main points and the logic of the answers provided by the participants.
- Be observant of inconsistencies in the participants' answers.
- Note any supplementary non-verbal communication.

5 Record responses

- Record the interview to prevent the need to write down responses.
- Invite the participant to raise any uncovered subjects.
- Thank the participants for their involvement.

Figure 92 Interview process

FOCUS GROUPS

Focus groups consist of small groups of participants who are brought together by a moderator to explore attitudes and perceptions, feelings, and ideas about a specific topic. Ideally, they include about six to nine participants, which is a good number to allow a dynamic discussion and the sharing of views and opinions, without being unmanageable. But, in small-scale projects the number can be smaller. It is important to consider the length of focus groups to avoid participants losing interest if they are too long or not having enough time to express their ideas if they are too short. Anything between 45 and 90 minutes is a reasonable length.

Focus groups are valuable for research involving user-centred design where it is important to find out what users think about the design output and how users interact with it. More importantly, focus groups help us determine whether an information design output is needed, or indeed what information design output is needed for that specific target user. Focus groups can also provide a less intimidating context for participants who might feel less comfortable in a one-to-one interview. However, there is the risk that more vocal and confident focus group participants take over or influence others because they are more assertive. Therefore, the focus group moderator needs to make sure this does not happen.

Focus groups differ from interviews because there is a focus/theme for the discussion and participants are recruited because they are knowledgeable about the topic. The information is obtained through group dynamics and interaction, which brings a different insight to inform the project because it reveals what people think, as well as why they think the way they do. In terms of moderation, the moderator facilitates the discussion instead of leading it, as is the case in interviews.

The success of a focus group also depends on trust. It is the moderator's responsibility to create an environment in which participants feel comfortable with each other to express their ideas. Participants also need to feel assured that personal feelings or matters will be treated as confidential and will not be disclosed by other participants in the group.

All in all, while facts give more objective data than opinions, it is nevertheless important to understand what motivates users, what their expectations are and what they perceive as being their needs. Getting this knowledge and considering it in the design of information is important. It gives us evidence of why and what information stands out among the vast amounts of information users are exposed to every day, and what also keeps users engaged for as long as needed.

HOW TO CONDUCT FOCUS GROUPS?

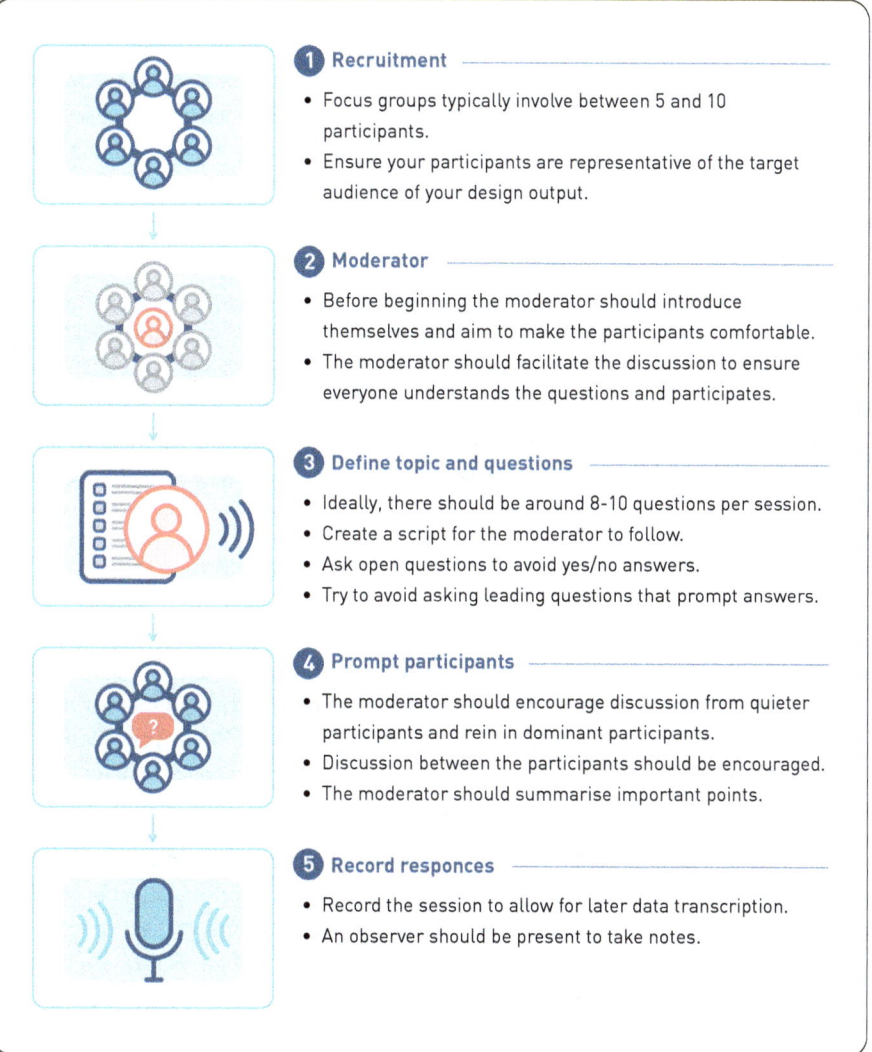

1 Recruitment

- Focus groups typically involve between 5 and 10 participants.
- Ensure your participants are representative of the target audience of your design output.

2 Moderator

- Before beginning the moderator should introduce themselves and aim to make the participants comfortable.
- The moderator should facilitate the discussion to ensure everyone understands the questions and participates.

3 Define topic and questions

- Ideally, there should be around 8-10 questions per session.
- Create a script for the moderator to follow.
- Ask open questions to avoid yes/no answers.
- Try to avoid asking leading questions that prompt answers.

4 Prompt participants

- The moderator should encourage discussion from quieter participants and rein in dominant participants.
- Discussion between the participants should be encouraged.
- The moderator should summarise important points.

5 Record responces

- Record the session to allow for later data transcription.
- An observer should be present to take notes.

Figure 93   Focus group process

VISUAL SURVEY AND HEURISTIC EVALUATION

Every information design project, including information visualisation, should start with what I refer to as 'visual survey and heuristic evaluation' (borrowing terms from other areas of design and arts). However, there is no fast rule, and it might be useful for certain projects to repeat the method at a later stage of the research and design process in order to update the findings if other features are worth looking at.

This method is used to audit and compile as many examples as possible of other work previously developed in the same area of focus. For example, if we are looking at patient information on how to follow a healthy diet and lifestyle while recovering from bowel surgery, then we collect as many examples as possible focusing on the communication of that particular information. If that information is to be provided as a booklet, for example, then we collect as many booklets as possible. If the booklets are to be used by patients only in the UK, then we collect as many booklets as possible in hospital and clinics in the UK.

A survey is then conducted where the objective is to look at and analyse how the information is visually presented. Hence the term 'visual survey'. A table/spreadsheet for this analysis is created based on well-established principles of information design (which will include information visualisation). Hence the term 'heuristic evaluation'. We then go through each example and check how they score against each criterion in the table/spreadsheet. For example: What type of font is used – serif or san serif? How many colours are used? What types of visuals (icons, graphics, illustrations, images, etc.)? And so on.

Naturally, this requires good knowledge of, or access to, well-established and research-based heuristics/principles of information design and visualisation. That is, principles that are determined by empirical studies and findings, by best practices in combination with empirical findings, or by best practices on their own where there is no research evidence, and by standards and conventions that have been observed and implemented over long periods of time. Researchers should identify these principles by always checking what has been published and what the latest findings are. Designers and practitioners who are more limited in time should look for reliable sources where these principles are listed, digested and explained to them (as in the case of this book and other sources identified in this book).

This will allow us to identify the problems with existing information outputs that are not following the principles, and where improvements can be made, as well as good practice that can continue to be applied. Moreover, if our intention is to conduct task analysis during the 'Discovery' stage, then this method allows us to identify possible examples for comparison in terms of good and poor design.

While the main focus is always on existing materials in the public domain that relate as closely as possible to what is going to be developed, there is also great benefit in bringing to this stage of the research and design process additional materials for analysis. This is especially the case when there are very few materials available in our area of focus and we need to find inspiration and examples of good practice from other related areas that can inform the development of our specific design solutions.

HOW TO CONDUCT VISUAL SURVEY & HEURISTIC EVALUATION?

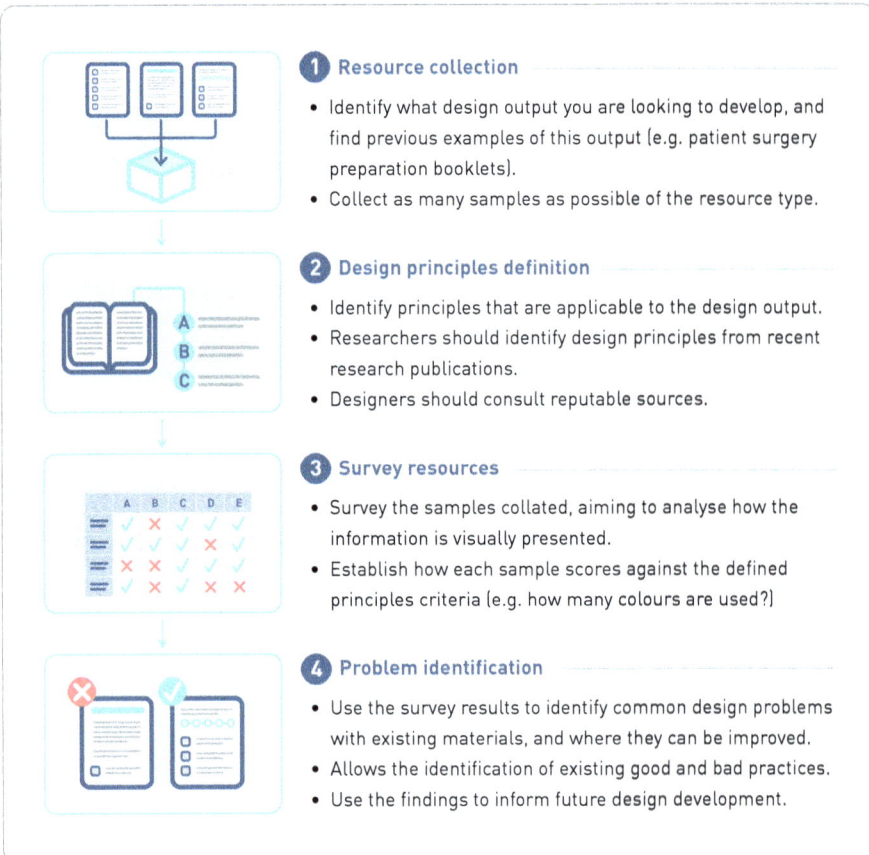

1 Resource collection

- Identify what design output you are looking to develop, and find previous examples of this output (e.g. patient surgery preparation booklets).
- Collect as many samples as possible of the resource type.

2 Design principles definition

- Identify principles that are applicable to the design output.
- Researchers should identify design principles from recent research publications.
- Designers should consult reputable sources.

3 Survey resources

- Survey the samples collated, aiming to analyse how the information is visually presented.
- Establish how each sample scores against the defined principles criteria (e.g. how many colours are used?)

4 Problem identification

- Use the survey results to identify common design problems with existing materials, and where they can be improved.
- Allows the identification of existing good and bad practices.
- Use the findings to inform future design development.

Figure 94 Visual survey and heuristic evaluation process

TASK ANALYSIS AND OBSERVATION

This method is very much a usability test but without iteration. Basically, an existing information design output is given to the target users of the interaction with the objective of identifying further problems and/or refining and improving existing designs.

Comparing 'task analysis and observation' to focus groups, the latter are a less valid and accurate way of evaluating design usability. To evaluate how users interact with the information design outputs it is important to closely observe each user individually. Just listening to what users think can be misleading. Although focus groups can contribute to the overall picture and have the benefit of generating discussion, it is very important to also watch how users actually interact with the design. Focus groups are, however, what helps us determine in the first instance whether a product or service is needed and solves an actual problem.

During 'task analysis and observation', target users are observed in action to better understand how they access information, meet their information needs, and interact with the information design output. It also gives insight at the cognitive level, i.e., how users act in terms of information processing, attention, memory and recall, problem solving, decision making, and so on.

All in all, it is useful to understand 'task analysis and observation' right at the start of the research and design process: what the objectives of the users are when accessing information; why they are accessing information; what will they do to fulfil their objectives; whether their previous knowledge, as well as the physical environment and platform they are using, influence the way they access information; what exact steps will they follow to perform the task in hand.

This method can be used to complement the findings from a 'visual survey and heuristic evaluation' once completed, or any other method during the 'Discovery' stage. Whatever method it complements, it should never be used in isolation to avoid running the risk of the facilitator/designer making their own interpretations of the problem instead of always having the needs of the user at the core of the analysis and observation. Moreover, like most other methods in user-centred design research, 'task analysis and observation' can be repeated if needed at a later stage of the research and design process in order to refine or extend the initial findings.

HOW TO CONDUCT TASK ANALYSIS AND EVALUATION?

1 Task identification

- First, identify the core task that needs to be analysed (e.g. asking the user to find medical treatment options on an app you have developed).
- Divide this task into 4-8 subtasks with specific objectives.

2 Recruitment

- Recruit participants that represent the target audience of your design output.

3 Task observation

- Create a diagram and written instructions with your 4-8 tasks that are presented to the participant.
- Ask the participant to complete the tasks and observe their behaviour and note any difficulties they have.

4 Problem identification

- The analysis should reflect the design's intended use.
- Difficulty with certain tasks will reveal areas of the design that need to be addressed.

Figure 95 Task analysis and evaluation process

EYE-TRACKING

Eye-tracking is a sensor technology that can follow and record what users are looking at in real time, i.e., visual attention. Because visual attention is directly linked to cognitive processing, as discussed in Chapter 1, measuring eye movements will reveal how users are processing information: are users easily finding and interacting with information, or are they struggling to find and read the information?

Eye-tracking is seen as an expensive and complex method to set up. I remember, when doing my first eye-tracking study as an early career researcher, how time consuming it was to set up the equipment. It was the same with setting the calibration and making sure participants did not move their head (which was placed on a chin rest) during the test, as this would compromise the results. Above all, I remember the anticipation I felt until the end of testing that the calibration had not failed and that eye movements were accurately recorded. The other problem was when participants arrived for the test and we realised that their eye colour and pupils were too bright for the calibration to work.

However, advances in technology now allow the use of eye-tracking at a much lower cost, the setting up of the equipment at a faster pace, and the success of the testing at a higher rate. Moreover, in addition to eye-tracking on a computer screen, eye-tracking solutions can now also be built on various other devices (e.g., eye-tracking glasses, laptops, tablets, virtual reality (VR) and augmented reality (AR) headsets), as well as bespoke systems in sectors such as education, healthcare, etc. This also means that eye-tracking has moved from being almost exclusively used in academy and research labs to being used commercially. Equally important, technological advances in eye-tracking allow a wider recruitment of participants and setting up the testing in many more environments. For example, we can include a wider variation of eye shapes, eye colour, and retinal reflectivity; there is more tolerance for the ambient light where the testing is being conducted; and there is also more tolerance for user movement while being tested.

WHAT DO WE ANALYSE WITH EYE-TRACKING?

We can analyse **fixations**, i.e., when users' eyes stop to take in detailed information about what they are looking at. A longer time to make a first fixation on a target piece of information means that the information is not standing out, or is not legible, or is not attractive enough. If, for example, the information is not legible or visually too complex, but readers are trying to read it, the average fixation duration on the target information will be higher because users are making a greater effort to process the information. If, on the other hand, for example, the information is clear and attractive, a longer average fixation duration can indicate that the information is accessible and engaging and users are taking their time to interact with it. This is a typical example where both quantitative (eye measure) and qualitative data are important to inform each other. That is, how do we know for sure whether a longer fixation is because users are struggling or engaging with the information? This can be clarified by collecting qualitative data at the end of the eye-tracking test and by interviewing users.

In addition to fixations, we can also measure saccades. **Saccades** are when the eyes move rapidly from one point of interest to another. Therefore, saccades indicate where users' attention is being directed, i.e., the areas of interest. When reading text, users only make small movements, whereas when looking at information on a page (an infographic, for example), eye movements are much larger. During each saccade users are unable to see, which means that we can only perceive and process information through successive fixations.

HOW IS EYE-TRACKING DATA COMMUNICATED?

Gaze plots show where, in what order, and for how long participants looked at the design stimulus. Gaze plots will therefore show the time sequence of looking from place 1 to place 2, to place 3 and so on, on an infographic, for example. Gaze plots also display the fixation duration, which is shown by the diameter of the fixation circles, i.e., the longer the user looks, the larger the circle is.

Heat maps show how what participants are looking at is distributed over the design stimulus. While the heat map only reveals the focus of visual attention, it can do it for one participant only, or for all participants at one time, regardless of the number of participants. Heat maps, however, can be misused and wrongly interpreted. Therefore, it is imperative to have in mind the aim of the study when constructing a heat map. What are we trying to find out and what is the data telling us?

All in all, I would say that gaze plots are better for visualising and communicating data on individual user eye movements and visual attention, and heat maps are better for visualising data on users as a group. Moreover, while heat maps can also reveal good data about individual eye movement patterns, doing the same with gaze plots will only result in an overwhelming number of circles overlapping each other.

For usability testing checking individual participant eye movements, the recommendation is to test eight to 12 users. In eye-tracking we need to account for calibration problems and other technical issues that might occur with eye-tracking. Therefore, we should test a higher number of participants than the five recommended for standard usability testing. In addition to a static visualisation, the eye movements of individual participants can also be analysed by watching the video recording of the eye-tracking experiment, which shows the gaze plots as participants' eyes move on the page. So, in this case, 8–12 participants is the ideal number, otherwise the analysis of the videos would take a very long time. However, when wanting to analyse a group of users through heat map, then the recommendation is to test around 30–40 participants, as quantitative analysis will be needed. Such a number also allows us to confidently do a statistical data analysis to identify whether there is a statistically significant impact or difference between designs.

Finally, just to highlight that although I am discussing eye-tracking at the 'Discovery' stage of my research and design process model, eye-tracking measurement is another method that can be used at other stages. For example, eye-tracking measurement is often used to validate a final design solution at the 'Evaluation' stage.

Gaze plot

Heat map

Figure 96   Eye-tracking visualisations

HOW TO CONDUCT EYE-TRACKING?

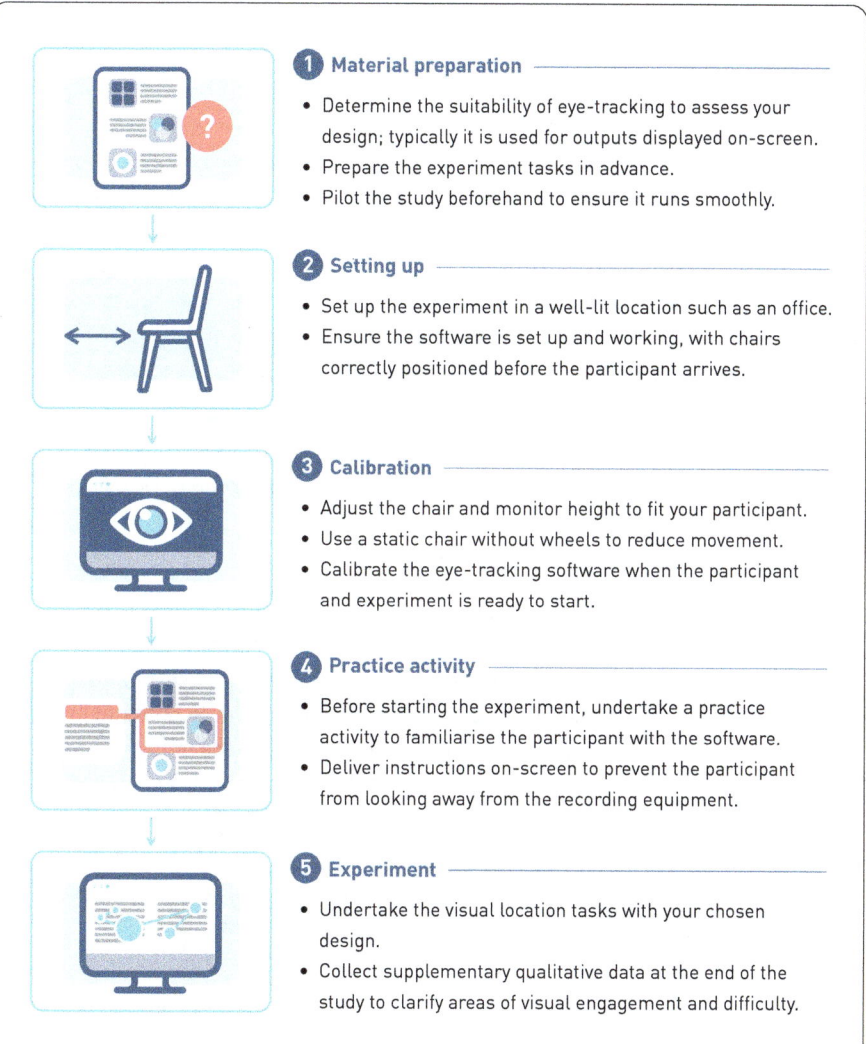

① Material preparation

- Determine the suitability of eye-tracking to assess your design; typically it is used for outputs displayed on-screen.
- Prepare the experiment tasks in advance.
- Pilot the study beforehand to ensure it runs smoothly.

② Setting up

- Set up the experiment in a well-lit location such as an office.
- Ensure the software is set up and working, with chairs correctly positioned before the participant arrives.

③ Calibration

- Adjust the chair and monitor height to fit your participant.
- Use a static chair without wheels to reduce movement.
- Calibrate the eye-tracking software when the participant and experiment is ready to start.

④ Practice activity

- Before starting the experiment, undertake a practice activity to familiarise the participant with the software.
- Deliver instructions on-screen to prevent the participant from looking away from the recording equipment.

⑤ Experiment

- Undertake the visual location tasks with your chosen design.
- Collect supplementary qualitative data at the end of the study to clarify areas of visual engagement and difficulty.

Figure 97 Eye-tracking process

DIARY STUDIES

Diary studies are used to log qualitative data over a period of time to give insight about how behaviour, needs, experiences and activities happen in real time. It would be too expensive and time consuming to replicate real scenarios that occur over an extended period of time in a lab setting. Therefore, giving participants a framework in which to log information about the activities being researched at specific times, and about particular things, is far more efficient and less time consuming. However, diary studies are not appropriate for all contexts, for example, when users are not physically able to record their activities because of the task itself (e.g., a patient recovering from surgery during the days immediately following the surgery).

When it comes to information design, diary studies are useful to define: (1) usage scenarios, i.e., when, how, and in what capacity do users engage with specific information outputs; (2) user motivations, i.e., what motivates users to access or not certain information outputs; and (3) information retention, i.e., how able are users to retain the information and act on it over a period of time.

The downside of diary studies, however, is that they are time consuming, in the sense that they are dependent on users acting and recording their actions across several days and even months. In addition, we have to trust that users will log everything accurately and not just what they feel comfortable with sharing. Finally, diary studies rely on user motivation and involvement that needs to be maintained for several days and not just during a 60-minute task analysis and observation.

Diary studies can bring a wealth of insight to user-centred research but should be used with purpose and to enrich the research and design process with a more qualitative approach, as well as with information that it would be impossible to collate in a lab environment.

HOW TO CONDUCT DIARY STUDIES?

1 Planning

- Define the aims and the timeline of the study.
- Generate the tools that the participants will use to record their diary entries, as well as detailed instructions.

2 Study briefing

- Meet the participants one-to-one to discuss the study.
- Emphasise the expectation and the scheduled time period that the study will run for.
- Explain the study materials and answer any questions.

3 Logging period

- Define clear instructions on activity logging using a simple framework, also providing log entry examples.
- Define when and how you require your participants to log a diary entry.

4 Interview

- After evaluating the diary entries, set up a one-to-one interview with each participant.
- Ask questions to uncover more details where necessary.

Figure 98   Diary study process

4.2 | EXPLORATION STAGE

Co-design

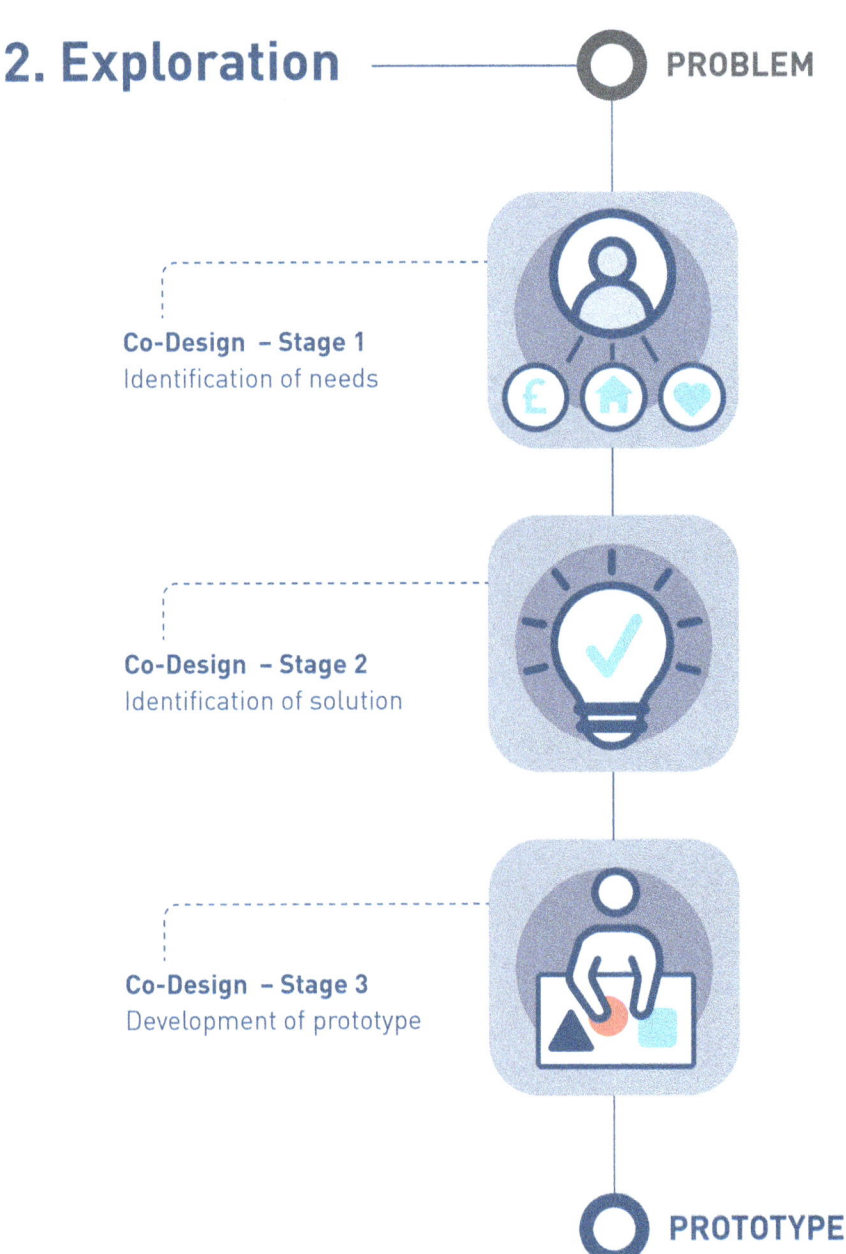

2. Exploration — PROBLEM

Co-Design – Stage 1
Identification of needs

Co-Design – Stage 2
Identification of solution

Co-Design – Stage 3
Development of prototype

PROTOTYPE

Figure 99 Exploration stage diagram

CO-DESIGN

Co-design offers an opportunity for designers/researchers to work with the end users of their information design output to understand users better, as well as to co-create to find the ideal solutions using creative activities. Co-design is collaborative at its core. This co-creation includes at least one designer and all relevant stakeholders, i.e., the users and, for example, public services representatives, industry and business representatives, etc. Designers and stakeholders work together to further identify the needs and find solutions to solve problems.

Please note that for the benefit of clarity, I will use 'researcher' to identify the person conducting and moderating the co-design session because in co-design a designer is considered to be one of the participants in the session. Even if the researcher is a design expert, when conducting the co-design session, their role should be limited to acting as the moderator and facilitator. Therefore, the researcher runs the session and participants either undertake the role of designer or stakeholder. The designers are the visual experts with professional know-how, who provide design guidance and creative thinking strategies for the activities. They are also the visual translators for participants who may not be skilled or confident in idea expression. The stakeholders are the users or parties who will invest or make use of the output being developed.

Co-design aims to make design accessible to everyone. It looks at value for the user and how to design for the user and with the user through a range of methods and tools. By giving users a direct voice early in the design development through co-design, we are increasing the usability and validity of the design solution. Moreover, despite creative activities being used, it does not matter whether those taking part have technical and creative skills or not. All can participate equally and describe and discuss their personal experiences freely.

Co-design should be used when the objective is to gain a better understanding of how people think about a particular matter, product or service. Co-design is a good method to adopt in contexts that are hard to observe, that are expensive to access, that are culturally sensitive, that are private, or even that do not occur very often. We often say that 'what users say they do' differs 'from what they actually do'. Therefore, building another opportunity to give users a voice, and this time in a context of teamwork and creativity, can help overcome that barrier and contamination of findings.

The co-design process can be divided into three main stages. In a way, it is a mini research and design process inside the main and overall research and design process.

STAGE 1: IDENTIFICATION OF NEEDS. In Stage 1, the researcher/moderator meets with users and starts familiarising himself/herself with the users, their needs and expectations. At this stage, a good co-design technique to employ are '**scenarios**'. Scenarios are used to describe the interaction between the product or service and the user, but from a user perspective. This description needs to be as simple and as clear as possible. The scenario technique is a good strategy to help focus on user requirements, instead of focusing on the less relevant technical or business requirements. Scenarios can be set up using prior knowledge from other research studies, for example. Another technique that can be used in Stage 1 is '**cultural probes**' to inspire participants to better understand and communicate about their lives, thoughts, environment and interactions. To fulfil their objective, cultural probes need to be imaginative and designed in such a way that are capable of evoking relevant responses and interaction from participants (e.g., text and imagery in general, magazines, newspapers, postcards, maps, cameras, and other recording devices). This technique is useful when the objective is to identify patterns and themes.

STAGE 2: IDENTIFICATION OF SOLUTION. In Stage 2, user needs, expectations, goals and values are determined and a desired focus and outcome is agreed. At this stage, a good research technique to use is '**card sorting**'. In the co-design session, participants are given cards with printed content, concepts, headings, terms and/or features and are asked to group items of information as they see best to enable them to easily find it and understand it. This is key to help identify information that might be difficult to find and navigate; to identify terminology that might be misunderstood; and to explore how participants group items into categories and relate concepts together. Card sorting is therefore useful for determining information structures, hierarchy of content and the content itself.

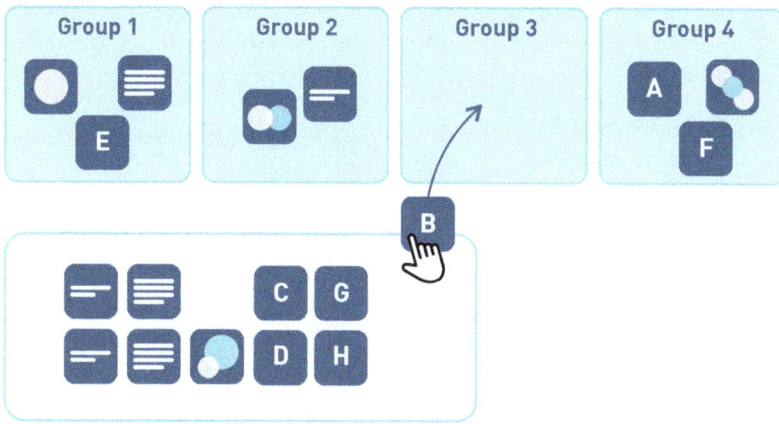

Figure 100   Card sorting

STAGE 3: DEVELOPMENT OF PROTOTYPE. In Stage 3, design solutions are developed to solve the problem and meet user needs and expectations. Bearing in mind that apart for one stakeholder profile (i.e., the design representative), no one else might have the creative and technical skills needed to develop a design outcome, at this stage, a '**creative toolkit and visual stimuli**' are used. This technique should include a range of assets and images that allow users to express various emotional reactions, such as characters expressing emotions (e.g., emojis with various expressions), icons (e.g., representing objects) and shapes (e.g., squares, circles), images (e.g., images of spaces and surroundings), textures, materials (e.g., wood, metal), assembly parts related to the design focus and platform or medium (e.g., for a website the toolkit should include interface buttons, menus, links), and creative raw materials (e.g., lego, colour pencils, clay). Another good research tool is '**storyboard**' as it enables users to illustrate the interaction that should happen between them and the product or service; or in the case of motion graphics, the actual content of the motion graphics. To achieve the best results, here too there should be iteration to allow a staged process of creation, reflection and evaluation. That is, participants work together to ideate (i.e., generate new ideas) but also to review and offer further solutions. All in all, Stage 3 is what we call in a design context, '**design workshops**'. Design workshops are a fun and engaging way of challenging participants to think creatively through activity-based research.

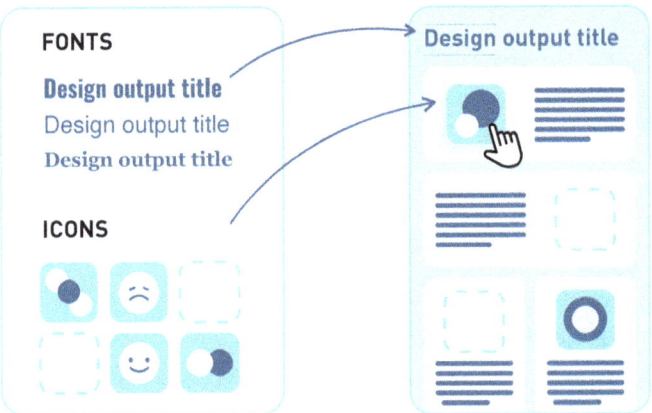

Figure 101   Design toolkit

There are situations, however, where stakeholders need to be grouped according to a specific role and responsibility and consequently a specific co-design session needs to be allocated to each group. This is because in certain contexts there is a default hierarchy among stakeholders that inhibits those in the low levels of the chain to freely express themselves, for example, to freely express their needs, their creativity and their problem-solving insights without feeling the pressure to compromise or feeling uncomfortable about expressing ideas and solutions. This is what I defined in one of my research studies (Lonsdale et al., 2020, p. 137), also explored as a case study in Chapter 5, as the 'in-between co-design' approach and as the 'within co-design' approach. An '**in-between co-design**' approach relates to the 'collaboration and sharing of ideas among a group of users with a common goal, and in the same process, but with different roles, with different needs and dealing with different angles of a problem' (Lonsdale et al., 2020, p. 137). This is the typical module used in co-design. A '**within co-design**' relates to the 'collaboration and sharing of ideas among a group of users with a common goal, with similar roles in the process, with the same needs and dealing with the same problems. All groups should always have at least one designer to assist in interpreting design tasks and design language' (Lonsdale et al., 2020, p. 137). The 'within co-design approach' happens when we divide stakeholders in groups to avoid inhibiting participants due to a natural hierarchy in roles, where each group only includes those with the same common goals, roles and needs (e.g., group 1 could include patients; group 2 could include nurses; and group 3 could include surgeons – see Case Study 4, Surgery Recovery, in Chapter 5).

Of all the user-centred research methods, co-design is the one that best helps the researcher empathise with the user and see the world from the users' perspective; that best creates an opportunity for multiple stakeholders to work together for the greater good of all in the team; that best makes the decision-making process accessible to all relevant parties; and that best brings individual views into focus through discussion but also through making and building.

HOW TO CONDUCT CO-DESIGN?

1 Toolkit development

- Before a co-design session, a creative toolkit and visual stimuli must be created.
- This aims to provide the user, despite of their design experience, the tools to create a suitable design output.

2 Identification of needs

- The researcher meets the participants, aiming to identify the needs and expectations of the target user.
- Techniques such as 'scenarios' and cultural probes' should be utilised to establish how to design for user.

3 Identification of solution

- The user requirements and values are used to propose a design solution that would address the users' needs.
- 'Card sorting' can be used to understand how the users structure the design content.

4 Design workshops

- The aim of workshop is to allow the user to create a design solution that solves a problem and addresses their needs.
- Users should be provided with a toolkit that allows them to generate this solution.

5 Refining the output

- Analyse your findings by removing information that is irrelevant to objective and summarising what was learned.
- Based on the ability of the user and the desired output, a specialist may be required to refine the final design.

Figure 102 Co-design process

4.3 | DEVELOPMENT STAGE

Usability testing

3. Development

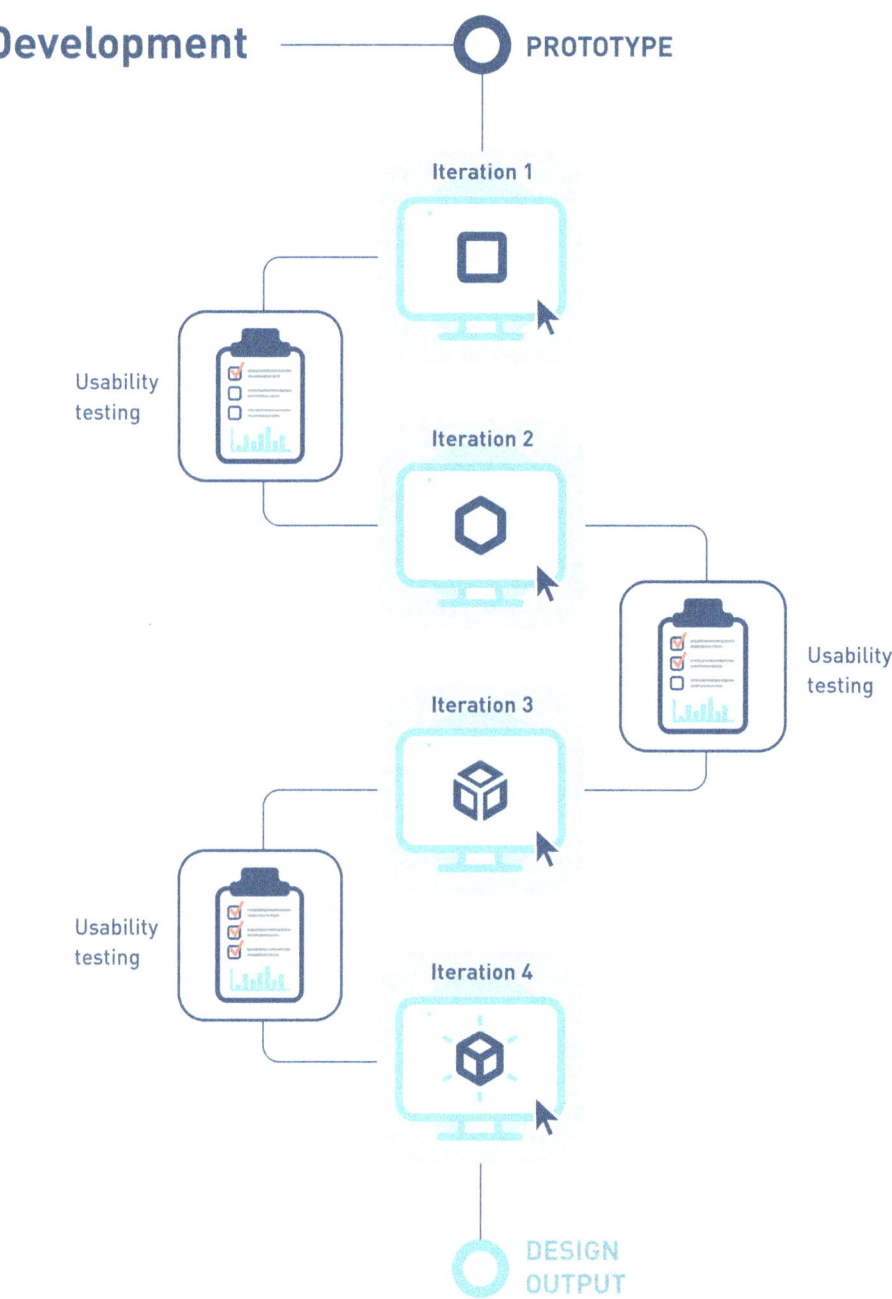

PROTOTYPE

Iteration 1

Usability
testing

Iteration 2

Usability
testing

Iteration 3

Usability
testing

Iteration 4

DESIGN
OUTPUT

Figure 103   Development stage diagram

USABILITY TESTING

After identifying the specific type of information design product/service that fulfils user needs and solves the problems in hand, it is time to find out if it performs to users' expectations. That is, if the design output is usable by that particular target user for that particular goal. As discussed throughout this book, accessing, navigating, digesting and assimilating complex information is a real cognitive challenge for users.

Usability testing is usually defined by five quality components that I will adjust slightly to fit with information design: (1) **Learnability**: the ease of accessing information and/or performing basic tasks the first time; (2) **Efficiency**: the speed of accessing information and/or performing the task and the accuracy of the information found; (3) **Memorability**: the ease of remembering information and/or performing a task after a period of not doing it; (4) **Errors**: the number of errors made by users, the severity of those errors, and the ease of recovering from those errors (in information design this relates more to digital and interactive information design outputs); and (5) **Satisfaction**: the pleasure of using the design.

Figure 104   Usability testing quality components

Users should not spend too much time and effort trying to understand the information or task. Research shows that when faced with difficulty, users' instinct is to give up. This is more and more the case in a society where the amount of information we are exposed to on a daily basis is overwhelming. If one design product/service is confusing and not easy to access and navigate, then there are plenty of others that users can opt for instead. More worrying is if users do have to access sub-optimal information and there are no other alternatives. In these situations, because users fail to understand the information or cannot access it easily, they either give up, or dismiss it, or decide for themselves what to do and end up using the information inadequately. This will then have a negative impact and do more harm than good (e.g., preparing for a medical examination that needs to be done rigorously, and on the day of the examination being turned down because the preparation is inadequate).

As already discussed, usability testing can be used at several stages of the research and design process. At an early stage – **Discovery** – usability testing can be used to assess an existing design and find out what is good or bad about it (defined in this book as the method of 'task analysis and observation'). At this stage, an option (and a cheaper option) might be to test the designs of possible competitors that might have similar features to ours, for comparison.

At the second stage – **Exploration** – usability testing can be used as formative testing, using initial concepts that do not need to include every feature, to better understand the users' needs and expectations. The approach is usually informal and collaborative, where the facilitator can ask questions as users access the information design product/service, and can also ask for design ideas. Alternatively, users can think aloud to describe their actions and thoughts about the design as they interact with the prototype. It is also possible to ask for ideas.

Usability testing is most commonly used during the third stage of the research and design process – **Development** – in order to develop the information design product/service in a progressive manner through various stages of design, testing and iteration. The first usability test at the development stage can be the testing of the first concepts isolated, or comparison of the various concepts. After this, we can decide on one concept, or merge more than one concept, to continue developing, testing and iterating.

Validation usability testing is a cheaper and less time consuming alternative to experimental testing (like A/B testing discussed later in this chapter, or any other form of experimental/empirical testing), which compares different designs. This would take place at the final stage of the research and design process – **Evaluation**. Here the success of the information design product/service is determined by comparing usability to certain standards by gathering quantitative data. For example, the percentage of information users should be able to find and/or understand, the time within which users can find the information, and so on. Another opportunity to conduct usability testing is once the final design has been fully tested, implemented and is live, if resources permit. This is because some subtle usability problems might still exist and can still be quickly fixed.

When it comes to how many **participants** to use in usability testing, there is a lot of debate on the exact number of participants, which to me is unnecessary and only serves to complicate rather than simplify matters. From my own experience of having conducted and supervised many usability tests, the guidance is very simple. As advised by many other researchers and practitioners, there is no gain in having more than five participants per usability testing when there are various usability tests and iterations in that research and design process. It is more productive, cost-effective and beneficial to have fewer participants per iteration, but various iterations and as many as needed to get close to the optimum information design product/ service. In sum, it is better to have five users early in the development and then five throughout the various stages of the development, than 100 users only near the end.

In terms of **prototypes**, the rule of thumb, considering costs and time, should be to start by testing low-fidelity prototypes (e.g., paper prototypes, static frames, etc.) and then move gradually to high-fidelity prototypes (i.e., more real and/or more elaborated design outputs). Multiple iterations should be conducted to refine the design. At all times, and I cannot emphasise this enough, it is important to test representative users and representative tasks, as well as to test users individually. Equally important is, during the testing, to let users overcome any problems on their own and never be tempted to direct them, otherwise this will contaminate the findings.

REMOTE USABILITY TESTING

What was initially a need to move our research to online platforms due to the Covid-19 pandemic has resulted in the discovery of more cost-effective, inclusive and wide-reach research methods. Some remote research methods, using computer programs/software applications, were available but seldom used. As with everything else, humans are creatures of habit and more often than not, unless we are challenged to do something different, we will just continue to apply what we are most familiar with and what we know works.

Usability testing as a remote user-centred research method can act as the basis for other methods that could benefit from having a remote option. Remote usability testing does not require participants or the facilitator to travel, making this a very good solution for situations where: (a) in-person testing is not possible; (b) users are in various and distant locations; (c) time is limited; and (d) the budget is tight. All these also allow for better and more accurate recruitment of representative users.

Remote usability testing can be moderated or unmoderated. **Moderated** remote usability testing refers to when the facilitator and participants interact with each other virtually and via screen-sharing platforms. **Unmoderated** remote usability testing, on the other hand, refers to the approach where the facilitator does not intervene in any way and participants complete the study in their own space and time, recording the session for later analysis by the researcher/designer. Nowadays, there are various high-quality software applications available to be able to run remote, unmoderated usability testing.

Naturally, there are also risks with remote testing as there are with in-person testing, and while these are different in nature, they should nevertheless be considered. It is important to make sure that instructions, tasks, questions, etc. are fine-tuned and piloted as many times as necessary to avoid problems and tests being annulled. This is especially important for unmoderated, remote usability testing. Do not rely blindly on the technology and always test everything thoroughly before running the test. As a safety blanket, it is better to recruit more users than you would recruit for an in-person usability test, in case there are problems with the testing and some data has to be rejected. Above all, we need to be extremely organised and think of all possible scenarios. In a way, we need to make sure to test the 'usability' of our own remote usability test.

HOW TO CONDUCT USABILITY TESTING?

1 Generate prototype
- Create a prototype design solution to address your defined problem.
- This should be an early model of the final output, aiming to learn about your initial concepts before further development.

2 Create plan
- Define the purpose of the usability test and problems you may want to address.
- Define the location, time table, moderator, and data collection method of the test.

3 Recruit users
- Recruit the testing participants, it is recommended to have five users per usability testing stage.
- Ensure these users represent the intended target audience of the design output you are testing.

4 Design the tasks
- Consider your prototype and what you want your target audience to be able to achieve by using it.
- Define representative tasks that will evaluate the core objectives of your design output.

5 Run the test
- Set up and pilot your test beforehand.
- Ask the participant to complete the defined tasks. The moderator should investigate any difficulties they have.
- Collect final feedback at the end of the session.

6 Analyse results
- Identify common problems that were brought up during the testing process and address them in the design iteration.
- Depending on the stage of the process, you should test the new iteration again after implementing the changes.

Figure 105   Usability testing process

4.4 | EVALUATION STAGE

A/B testing

Desirability testing

4. Evaluation

DESIGN OUTPUT

Stage 1 – A/B Testing:
Quantitative data collection

Stage 2 – Desirability Testing:
Qualitative data collection

Stage 3:
Data analysis

SOLUTION

Figure 106   Evaluation stage diagram

A/B TESTING

A/B testing is the simplest experimental testing method to check which information design users perform better with, where two information designs – Design A and Design B – are compared. Experimental testing refers to empirical research where experiments are conducted with statistical analysis.

Other formats test three or more information designs. However, these will not be discussed here because it is very rare that four or more information design outputs are compared within the thorough research and design process that I am describing in this chapter. First, because by the time we get to the 'Evaluation' stage of the research and design process, we have already refined the information design solution to the point that we should not have many design outputs to test (we might have one, maximum two). Second, because comparing four different information design outputs through experimental testing and statistical analysis would make the test too long for participants to complete, which in turn might contaminate the results as users' attention span will decrease with the later information design outputs. So, as a rule of thumb, it is better not to test more than two information design outputs at this late stage of the research and design process. Moreover, one might just want to compare a new information design output against an old design to check whether the new version helps users to access information more easily and perform better overall. That is, when compared to the old version, how quickly is the information found and navigated through with the new design? How accurately is the information found? Is the information easily recalled at a later stage?

A/B testing only gathers quantitative data. Therefore, to collect qualitative data in an experimental study with statistical analysis, other methods should be included at the end of the performance test, such as user feedback and opinion, desirability testing (as explained next), and others that are able to give us data on users' needs, struggles with the design, feelings about the design, and so on.

When comparing designs through experimental testing with statistical analysis, it is also important to consider very carefully the way the experiment is designed and how different conditions are assigned. The 'design' of an experiment in this case refers to the way we plan and structure the experiment; it does not refer to the visual design of the outputs being tested. 'Conditions', on the other hand, refer to the different visual information design outputs. For illustration purposes, A/B testing has two conditions, where condition 1 can be the 'Old Design' of an infographic, for example, and condition 2 can be the 'New Design' of an infographic.

Failing to design an experiment properly can result in the data not being usable after all the time and money spent on the research. In this regard, we have two options to consider. **Within-subjects design**, where the same user tests both information design outputs. In this case, the information design outputs are randomly assigned, i.e., alternated so that participants do not start always with design A, but 50% of participants start with design A followed by design B, and 50% of participants start with design B followed by design A. This is so that we reduce order and learning effects and mitigate the risk of these effects contaminating the results. That is, if design B is always shown second, it is very likely that users will perform better with design B because they are more familiar with the content, or/and the design style, or/and the navigation process, etc.

Between-subjects design, on the other hand, means that one group of participants tests design A, and then another group of participants tests design B. In this case, twice as many participants need to be recruited. So, for example, if a minimum of 40 participants is needed to show statistical significance with confidence, then for a between-subjects design, we need to recruit 80 participants, i.e., 40 to test design A and 40 to test design B. This means that it is twice as expensive and twice as time consuming to conduct the testing. Moreover, we would need to consider a kind of purposive sampling, i.e., selective recruitment of participants, so that we have a similar sample for design A and design B (similar number of male and female participants, similar number of participants per age group, educational level, nationality, etc.)

The major advantages of a within-subjects design are that it is less time consuming and less expensive for the researcher/designer, and any problems between design A and B will not go undetected or be masked by the fact that we have two different groups of participants. The major advantage of a between-subjects design is that it is less time consuming for participants because they are only testing one information design, and, as mentioned above, there are no order or learning effects between one design output and the other.

It is important to note the difference between co-design and A/B testing when referring to within-subjects design and in-between subjects design.

HOW TO CONDUCT A/B TESTING?

1 Test material generation

- Decide which two designs you will be comparing.
- For example, comparing an old design with a new one that has been redeveloped through multiple iterations of usability testing.

2 Hypothesis generation

- Identify the problem that the design you are testing aims to address (e.g. instructional material for patient education).
- Define hypotheses based on informed predictions (e.g. design A will perform better than design B).

3 Design the experiment

- Determine the performance measures you will use to compare the two designs.
- Define qualitative data collection methods (e.g. interviews) to support and explore the quantitative data.

4 Recruit participants

- Decide if you are using a within-subject or between-subjects comparison method, as this will determine sample size.
- Typically, large sample sizes are required for accurate statistical comparison of quantitative data.

5 Run the test

- Pilot the test before starting it with two or three participants to check everything runs smoothly.
- Run the test and remain present to monitor the participants progress.

6 Analyse results

- Determine if the results supported or rejected your hypotheses, and if they were significant.
- If results were unexpected, explore potential reasons why.

Figure 107   A/B testing process

DESIRABILITY TESTING

When there is no clear statistical evidence in favour of a design against another, or in favour of improvements made to a design, desirability testing is a good complement to statistical analysis. Desirability testing helps us to understand what may lead to an overall positive or negative emotional response from users, or what might attract users (or not) to a particular information product/service in the first instance. As we discussed before, first impressions matter at the iconic memory stage because what users see in those first split seconds matters in attracting their attention; and at the short-term memory stage because the way information is designed and visualised will help users determine whether to send the information to the long-term memory and recall it at a later stage or dismiss it.

Desirability testing goes beyond the simple collation of user feedback and opinions. Instead, it helps us to understand how people react to different information design outputs and, therefore, does the design need to be adjusted if the emotional response we are seeking is different from what users are expressing? It also helps us to understand, when comparing designs, which one provokes the emotional reaction we seek, such as a first impression about a design, a desire to continue to be engaged, whether users would use it again in the future, and so on.

To this end, users are given a list of adjectives within two categories: positive adjectives and negative adjectives. A third category can be added with neutral adjectives. For the positive and negative adjectives, it is always good to give related adjectives. For example, if choosing 'easy', then give the opposite adjective 'difficult'. The adjectives should also be tailored to the particular task users are being asked to perform, and the information design product/service they are being asked to interact with. Users need to be able to express their emotions and tell the story through these adjectives. To put it in a simple way, we show the design to the user, ask them to interact with it by performing a task, and then ask them to select three, four or five adjectives that describe how they feel about the design.

Desirability testing can be used as a quantitative method, by calculating what adjectives and percentage was selected for each design. It can also be used as a qualitative method by asking participants to explain each adjective selected. My suggestion is that if it is used as part of usability testing and iteration with only about five participants at a time, then it is important to have qualitative data as well. If used as part of an A/B testing or other empirical testing, then quantitative analysis would be sufficient, with an overall qualitative question at the end asking participants to briefly explain their choice of words. Otherwise, the test might become too long with other measures included. But, if time allows, then asking them to explain the choice of every adjective is beneficial.

HOW TO CONDUCT DESIRABILITY TESTING?

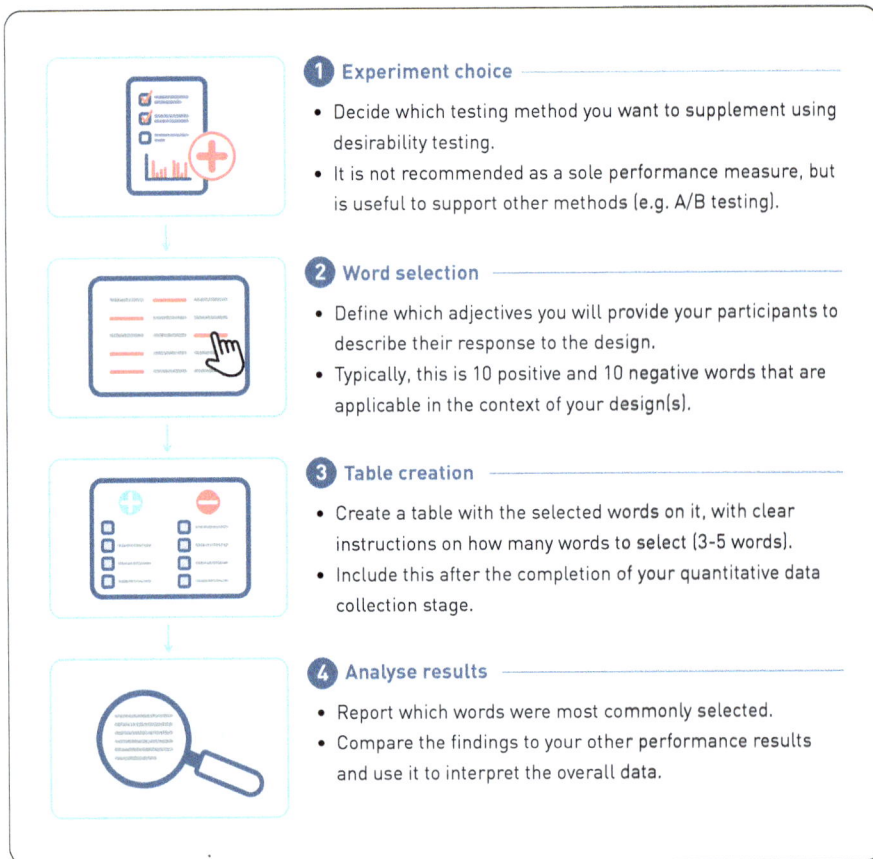

1 Experiment choice

- Decide which testing method you want to supplement using desirability testing.
- It is not recommended as a sole performance measure, but is useful to support other methods (e.g. A/B testing).

2 Word selection

- Define which adjectives you will provide your participants to describe their response to the design.
- Typically, this is 10 positive and 10 negative words that are applicable in the context of your design(s).

3 Table creation

- Create a table with the selected words on it, with clear instructions on how many words to select (3-5 words).
- Include this after the completion of your quantitative data collection stage.

4 Analyse results

- Report which words were most commonly selected.
- Compare the findings to your other performance results and use it to interpret the overall data.

Figure 108   Desirability testing process

OVERALL CONCLUSION

How many of us in our career as designers made the naive mistake of spending hours and days developing and fine-tuning the perfect design prototype for testing, being very confident that such a design is what the users need and it will work well with just some adjustments, only to find out after showing it to a couple of users that it does not work? I guess the answer is quite a few of us. This is one of the many reasons why research is imperative if we are serious about designing the optimum information output for those particular target users at that particular time in their lives.

Moreover, even after years of experience and knowing the inside out of good information design, every time we design a new output there are different variables to consider: the users are different, the information is different, the platform might be different or at least might have improved in its functions, and certainly time has passed, and things evolved. This means that every time we create a new design output, we always have to start the research and design process from scratch.

Finally, let us look at the way I divided the different research methods across the four stages of my research and design process model. The first stage – Discovery – has the highest number of methods of the four stages. This indicates that a reasonable amount of time and strategic planning needs to be dedicated at the start of the research and design process to identify the problem and user needs. It is important for us creatives to remember this because our instinct is to move quickly to what is most natural to us, i.e., creative design. The second stage – Exploration – and the third stage – Development – only have one method allocated to each one of them. However, these are in themselves a mini research and design process inside the main overall research and design process needed to devise the optimum design solution from beginning to end (as per my model). Then the fourth stage – Evaluation – includes two methods, but they should essentially always be conducted together in order to gather more comprehensive data through both quantitative and qualitative data collection.

WANT TO LEARN MORE ABOUT THIS TOPIC?

Martin, B. and Hanington, B. 2019. *Universal Methods of Design: 125 Ways to Research Complex Problems, Develop Innovative Ideas and Design Effective Solutions* **(expanded edn). Beverly, MA: Rockport.**

This is the book I always go to at the start of a research study just to remind myself again of the many research methods available and variants. The breath of design research methods listed (125 methods) helps us to really think about what methods we should use to answer our research questions and solve the design problems while meeting user needs. In addition, the content is very accessible and nicely layed out in no more than two pages per method: one page to explain the method and one page to exemplify the method with images and captions. What is also very valuable is the way the authors allocate numbers to each method from 1 to 5, to indicate at what stages in the research and design process such methods can/should be used.

Muratovski, G. 2022. *Research for Designers: A Guide for Methods and Practice* **(2nd edn). London: Sage.**

As well put by Gjoko Muratovski, craft skills no longer suffice to produce well-designed outputs and therefore we need to discover, define and solve problems based upon evidence, as well as demonstrate the validity of our claims. This is what this book is able to teach and equip us to do, i.e., it contains the knowledge and tools to conduct and make sense of design research throughout the design process.

Marsh, S. 2022. *User Research: Improve Product and Service Design and Enhance Your UX Experience* **(2nd edn). London: Kogan Page.**

Although this book focuses on user research for UX experience, it works perfectly to guide any user-centred research in the field of information design and information visualisation. Its content is very easy to digest and understand. Despite not having many images to illustrate the content, it has a good structure that allows us to quickly identify the various things to consider when conducting user research. For example, the content is structured around: 'what is the method'; 'when to use the method'; 'how to use the method', etc. Every method and respective section is also well supported with references at the end of the section.

Sanders, E.B.-N. and Stappers, P. 2022. *Convivial Toolbox: Generative Research for the Front End of Design.* **Amsterdam: BIS.**

As the title indicates, this book focuses on the first stages of the research and design process where generative research is produced. Its main focus is to equip readers with creative techniques and tools that allow researchers/designers to involve the users directly in the design process. As I also argue, Sanders and Stappers claim that design must be made with the people, not only for the people who are being served through design. This book has as many as 310 pages filled with creative and interesting methods and techniques, and all well supported with imagery.

Toptal Product Blog (www.toptal.com/product-managers/blog) and Nielsen Norman Group (www.nngroup.com)

These are two extraordinary online sources of information when it comes to explaining and illustrating user-centred design research methods. Toptal Product Blog is excellent at providing very insightful, down-to-earth, and up-to-date content in combination with excellent imagery and well-designed infographics. The Nielsen Norman Group has been established for years and is extremely reliable in providing insightful content based on research evidence, years of knowledge and always well supported with references from other credible authors.

5

From Theory, to Research,
to Practice... and Back

INTRODUCTION

Progress and technological advances have undoubtedly increased our access to knowledge and information. This benefits society in many ways, but it also means that information flow has massively increased. This is a challenge that information design and visualisation can help overcome by identifying and applying smart ways of reducing information load.

Infographics' popularity as a visual approach to communicate complex and dense messages has grown significantly in the last seven years (as evidenced in Google trends). This is due, for example, to the ability of infographics to encapsulate several details in one visual, while still being clear and precise.

However, despite this increased interest in the benefits of information visualisation, after a thorough review of the literature, it is evident to me that research studies on its effectiveness are very scarce. Beyond a few academic papers, the literature on infographic design and information visualisation is largely unscientific and limited to a few books providing recommendations targeted at business or journalism. Moreover, these recommendations are essentially practice-based, i.e., they derive from tacit knowledge acquired through practical experience. When scientific studies are unavailable or do not give clear answers, design practice can aid our understanding of how design principles can be applied to produce effective design solutions.

Ideally, we need empirical evidence to validate design solutions, to ensure that they fulfil their goal of clear and accessible communication with the target audience. However, I have also identified that the few research studies available are also limited in scope. For example, they test a limited number of visualisations, and often these are designed by the researchers without the involvement of a designer or in consultation with the user, which compromises the reliability of the findings. They also test a limited number of participants, some of whom are not always representative of the target user. In sum, there is clearly a great need for more studies on the effectiveness of infographics and data visualisation to communicate complex and/or large amounts of information. Moreover, how these have direct applicability to practice can make a positive impact on people's lives.

To meet this need, in this chapter I present six empirical studies conducted by me and my research and design teams, and in one case supervised by me. These studies show a diversity of areas where information design and visualisation can enhance user performance. These are studies that have already been published in blind peer-reviewed international journals, which denotes their great contribution to knowledge in the field of information design, as well as their impact beyond design in areas such as healthcare, security and intelligence, and education. But, while the academic papers focused on the research side, the case studies presented in this chapter focus more heavily on the creative side, i.e., design development, testing and iteration.

5.1 | CASE STUDY 1 |
CANCER SCREENING

SPECIFICATIONS

TITLE: Information design and visualisation for bowel cancer detection.

DESIGNER: Li-Chin Ni.

RESEARCHERS: Maria dos Santos Lonsdale, Maureen Twiddy, Li-Chin Ni and Chenyi Gu.

PUBLISHED IN: Lonsdale, M.DS., Ni, L.-C., Gu, C., and Twiddy, M. 2019. Information design for bowel cancer detection: the impact of using information visualisation to help patients prepare for colonoscopy screening. *Information Design Journal*, 25(2), 125–156.

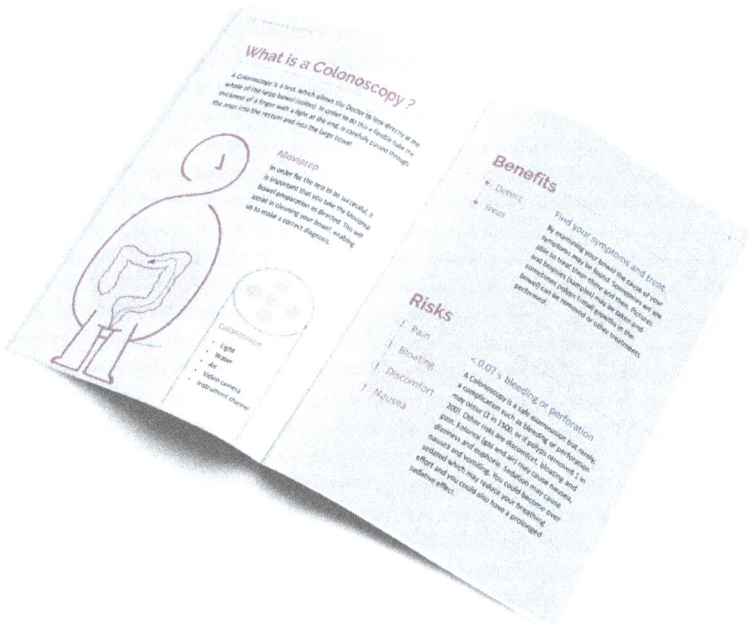

Figure 109   Final booklet designed for the Cancer Screening case study

ABSTRACT

PROBLEM/NEED: Colorectal cancer is one of the most commonly occurring cancers in the world, but is very treatable and most patients survive colorectal cancer if it is found early. Colonoscopy is the best procedure to detect it, with 95% sensitivity. However, unlike other health exams, colonoscopy success depends on the quality of bowel preparation and the percentage of poor bowel preparation is as high as 25%. An important factor in the quality of bowel preparation is the clarity of instructions, but the way information is designed and communicated to patients does not meet their needs.

AIM: To improve the ease of finding information and comprehension on how to prepare the bowel to undergo colonoscopy screening by designing, testing and finding the best design solution for bowel preparation instructions.

SCOPING RESEARCH: Literature review was conducted to understand the context and problems. A survey of existing booklets was conducted to determine a typical layout and content to be used as a baseline to improve the design of the instructions. A focus group with members of the public who have had a colonoscopy previously was conducted, followed by an online questionnaire to collect quantitative and qualitative data on patient understanding and perception of the design and content of the instruction booklets.

DESIGN DEVELOPMENT: Several usability tests were conducted to develop, test and iterate the outputs.

DESIGN OUTPUTS: Booklet and motion graphics.

USER PERFORMANCE TESTING: Experimental testing and statistical analysis were conducted to evaluate and validate the final outputs.

SCOPING RESEARCH

Through a **survey** we identified that the majority of the booklets explained the nature of a colonoscopy and how to take the purgative medicine (while the remaining booklets relied on the instructions provided with the medicine itself). The majority of the booklets also conformed to good typographic legibility principles and used less than three colours, with the main text being mostly presented in black. The biggest failing was the lack of visualisation.

In a **focus group**, four participants aged between 60 and 70 years old, who were former colonoscopy patients, were asked to choose one booklet from the ones surveyed. This booklet then served as the study material and was redesigned according to the principles of information design and visualisation, as well as participant feedback. Participants gave feedback for the design of the booklet as follows. They liked to see in one booklet a table with information on what to eat because they considered it was important for this information to be communicated very clearly. They also liked to see a diagram on colonoscopy. They further mentioned that a redesigned booklet should accommodate the different needs that different patients will have. For companion outputs, patients selected a companion video motion graphics, which would be more accessible and could be used directly by them, their carers and nurses.

In an **online questionnaire**, 32 participants over 50 years old, who were also former colonoscopy patients, were asked to list good and bad things about the design of the booklet they had received when referred for a colonoscopy. Participants were also asked about the design of the existing booklet in order to identify their expectations for the redesign, such as design features to consider and possible companion outputs. Main design features included: more visualisation of information using illustrations, infographics and tables; san serif typeface; avoid colour red, brown and yellow; use a friendly tone. The companion output that was most selected was again a motion graphics.

DEVELOPMENT, TESTING AND ITERATION

Literature was systematically reviewed to identify practice- and research-based design principles that also informed the design development of the outputs, in addition to all the primary research discussed in the previous section. An existing booklet was redesigned and a patient-support motion graphics was also created that could be used as a companion output to the booklet or to be used as stand-alone outputs. The development of the two outputs included various stages of design, testing and iteration.

BOOKLET

Participants selected an existing booklet to be redesigned, which was text dense (i.e., with no visuals), used two colours (purple for the headings and black for the main text), had a legible typographic layout, and used bold to emphasise important information. The development of the redesigned booklet involved three stages of design, testing and iteration.

- In **Stage 1**, four design and illustration styles were initially developed to identify characters, colour palettes, etc.
- In **Stage 2**, four low-fidelity prototypes for the patient-support booklet (four pages per prototype) were created after developing rough drafts. The prototypes contained a fair amount of visualised and logically structured information and included four colour palettes.
- In **Stage 3**, two prototypes were selected by the research team based on the results of the questionnaire with former patients and information design principles. The two prototypes were developed into high-fidelity prototypes and a detachable preparation schedule was also created to meet the request by participants during the focus group to provide a basic summary of what they need to do when preparing for the exam.

STAGE 1

Initial sketches

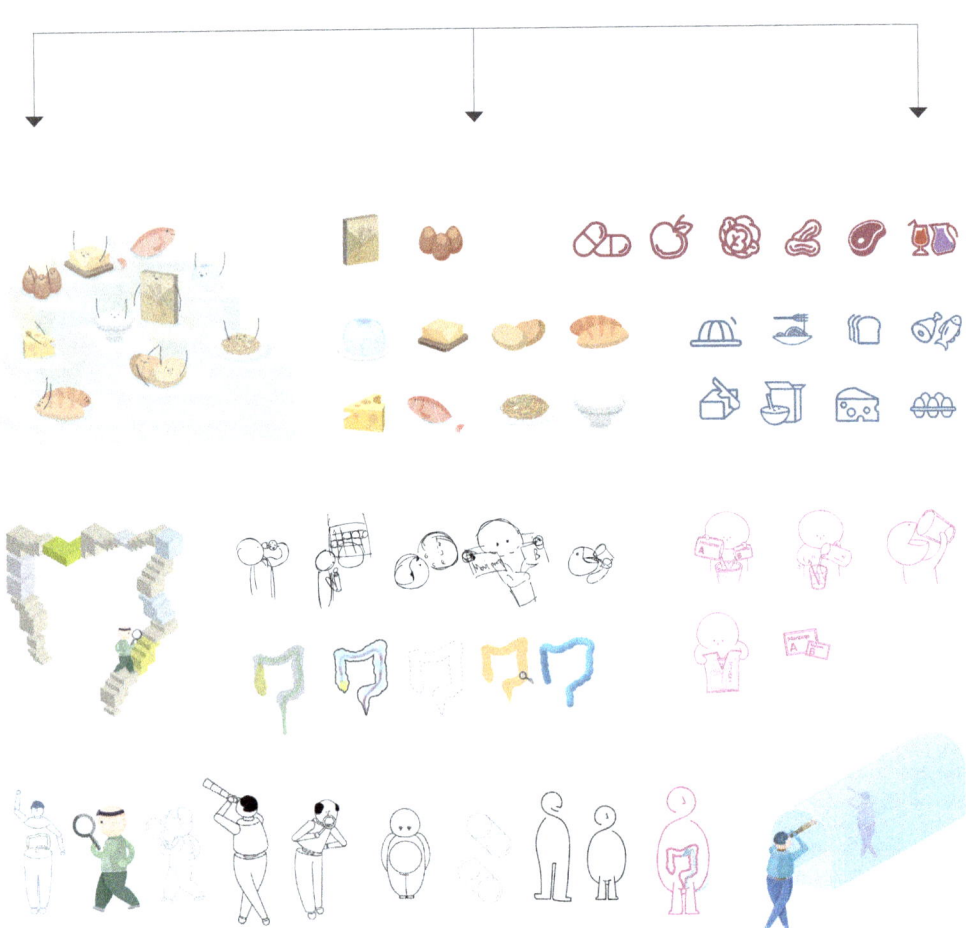

Figure 110 The initial sketches exploring illustration styles for the booklet

STAGE 2

Booklet . Concept A

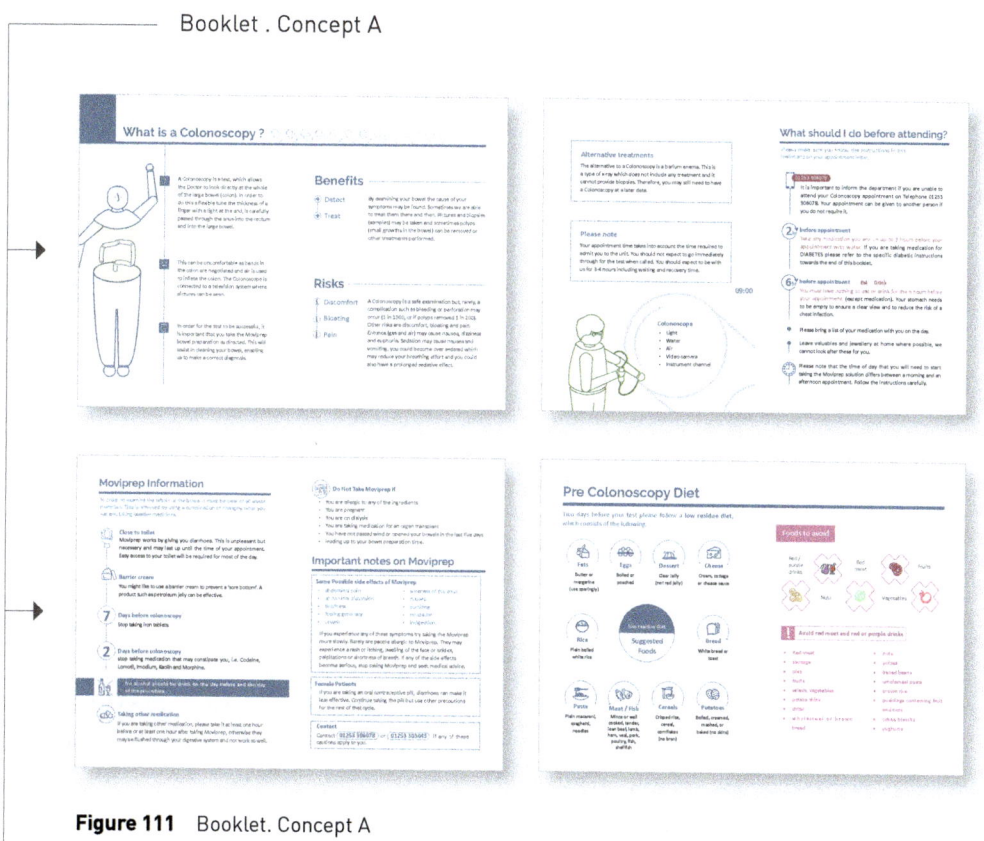

Figure 111 Booklet. Concept A

The focus of this concept was on maximum simplicity in terms of illustration and icon style and infographic design. Regarding the layout, the focus was on a well-defined and neat structure, maximum clarity in terms of hierarchy, and the reduction of text density. The colour palette was based on a neutral dark colour tone with the use of a warm colour for emphasis and contrast. In terms of typography, only one font was used and type size, type weight (e.g., bold) and colour were manipulated to emphasise and establish hierarchy.

STAGE 2

Booklet . Concept B

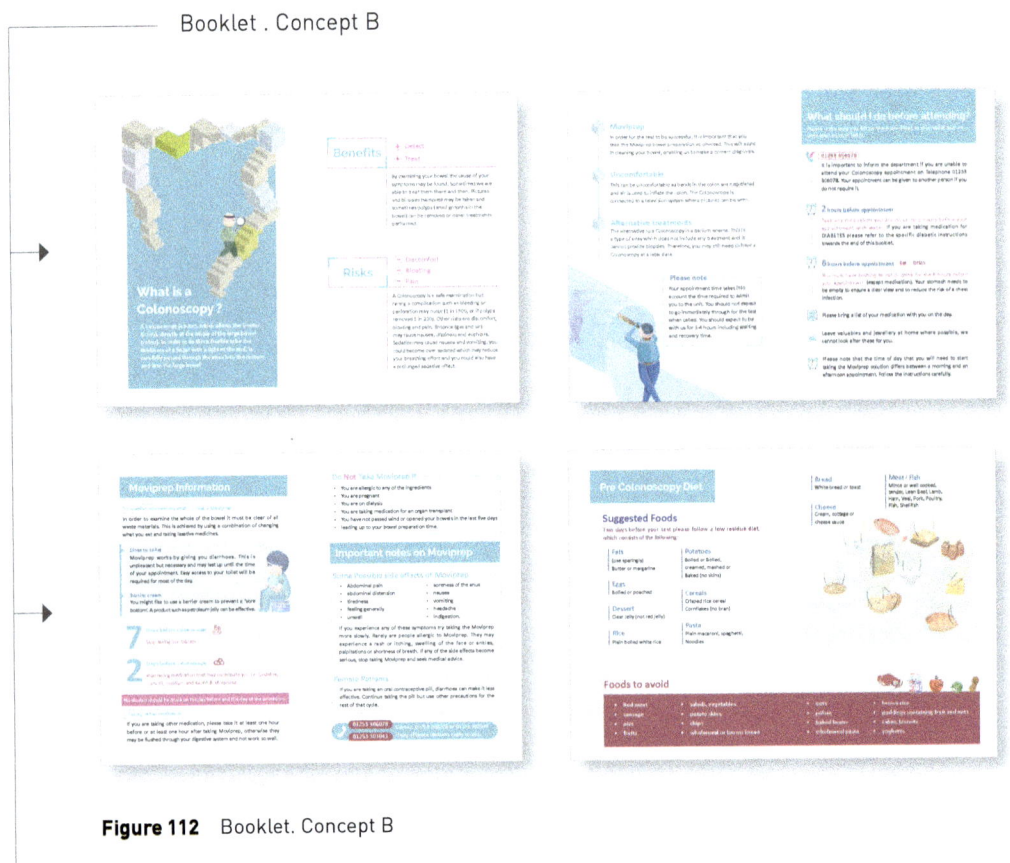

Figure 112   Booklet. Concept B

Concept B attempted to create a more friendly tone by using a softer colour palette with pastel colours. Illustrations were also more realistic and detailed for the same purpose and to create a bigger sense of familiarity.

STAGE 2

Booklet . Concept C

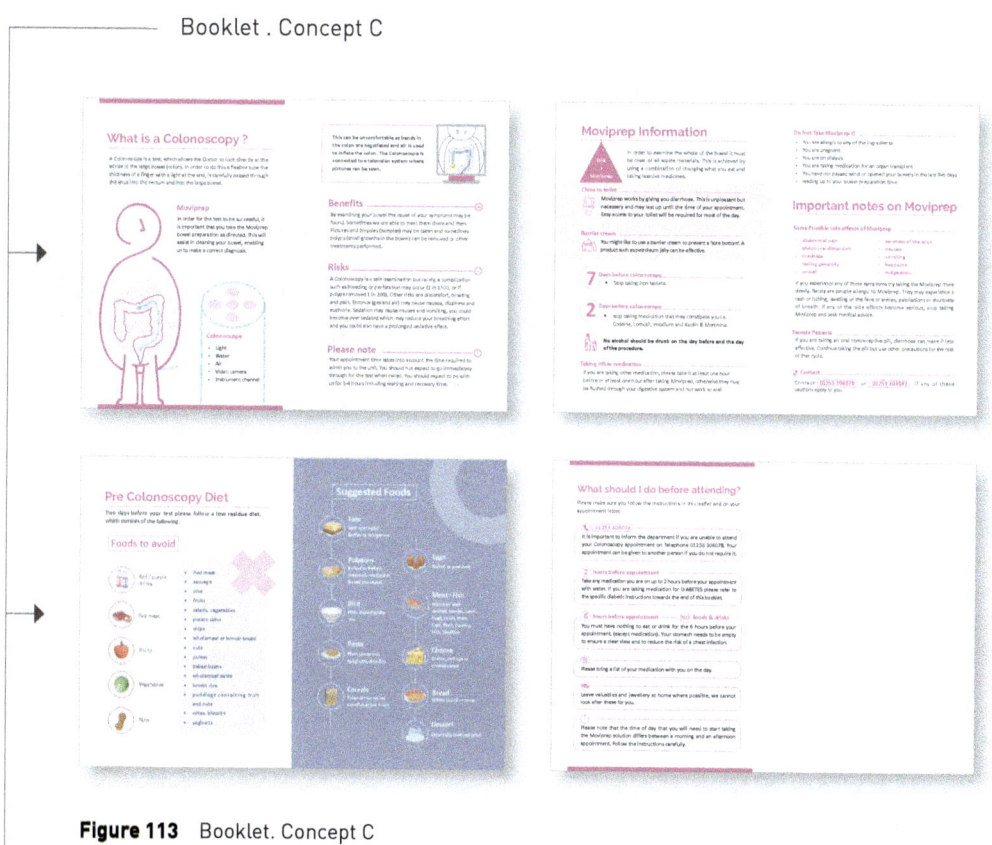

Figure 113 Booklet. Concept C

Concept C brought together the simple illustration style of Concept A and the more realistic and detailed illustration style from Concept B that was applied to the food. The same approach was followed as in the previous concepts to create a well-organised structure, a clear hierarchy, and the use of icons and illustrations to decrease test density. A warm, positive and bright colour palette was used.

STAGE 2

Booklet . Concept D

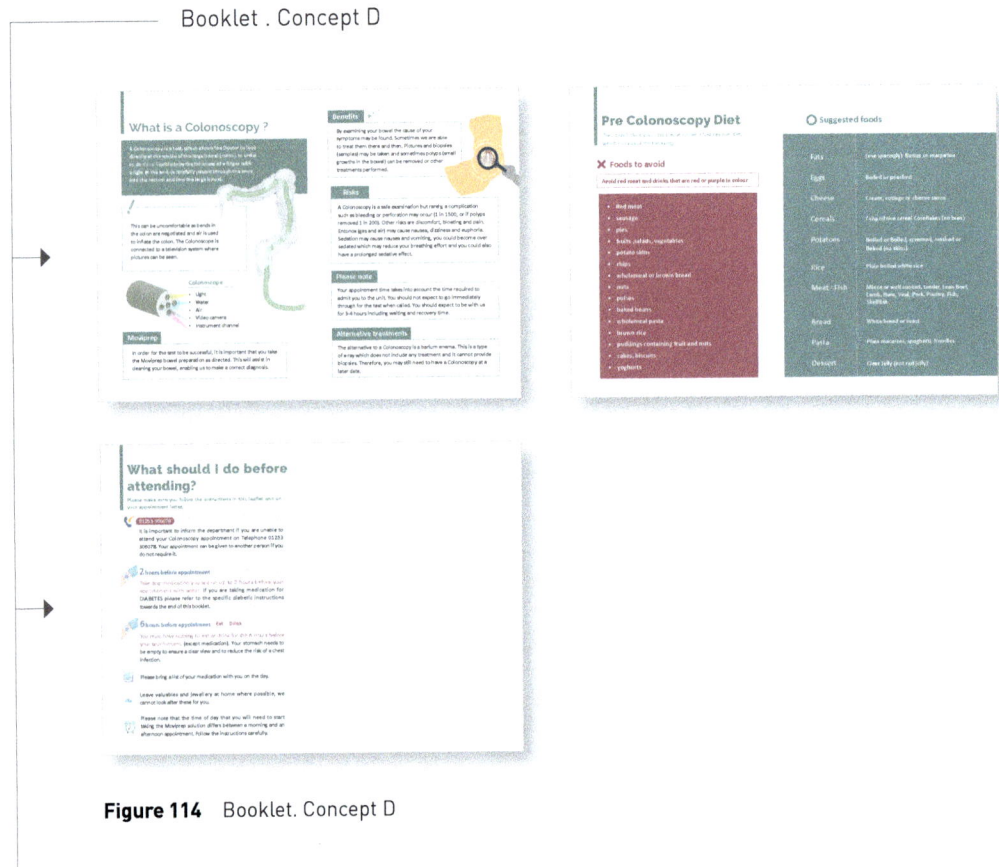

Figure 114 Booklet. Concept D

Concept D was designed to be more formal and corporate. This was achieved through a more rigid illustration style, boxes enclosing the headings and the darker colour palette.

STAGE 3

Schedule. Match Concept A   Schedule. Match Concept C

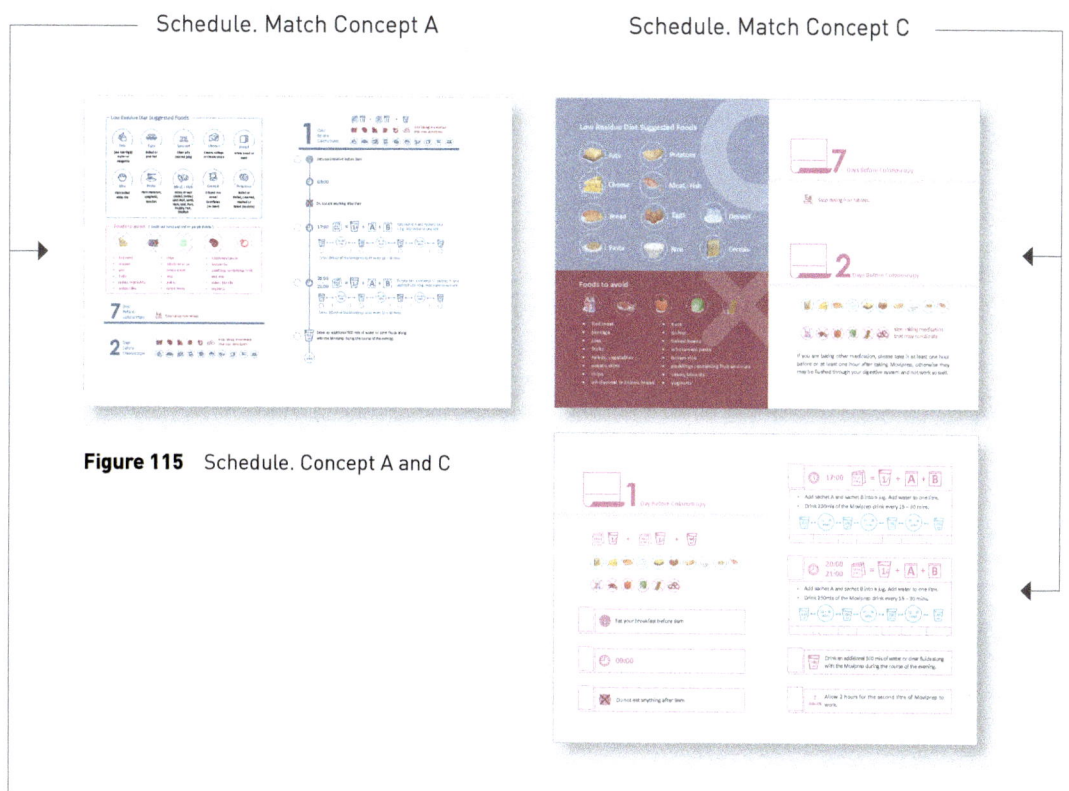

Figure 115   Schedule. Concept A and C

Concept A and Concept C of the booklet were selected and developed into high-fidelity prototypes. A detachable preparation schedule was also created to match each booklet, based on the feedback from participants during the focus group. In order to prepare the bowel for colonoscopy, patients need to follow a series of steps (stop taking some medicines, change diet, etc.) a few days before the exam. Having a schedule at a glance was a very useful item.

A usability test was conducted with six participants to test the two prototypes selected: four female and two male; two below 50 years old, four above 50 years old, with an average age of 48 years old. Results were very positive: all participants could find the correct pages and sections in a short period of time; five participants could briefly tell how to prepare for the exam; two participants could memorise the 'diet list'. In terms of opinion about the design of the booklet, both designs received mostly positive feedback, but Concept C received more positive feedback (76%) than Concept A (53%).

Overall, feedback for Concept C was as follows: (1) Infographics and illustrations – clear, easy to read, accessible, stand out, friendly, pleasant, modern, approachable, descriptive; (2) Colour palette and colour coding – friendly, nice, appropriate, secure, relaxed, fresh; (3) Detachable schedule – helpful, clever, clear, nice, appropriate size. Concept C booklet and schedule were developed fully, and further suggestions given by participants were taken into account. These included: to have a bit more colour than only one colour; to have more infographics; and to have more illustration or icons for key points.

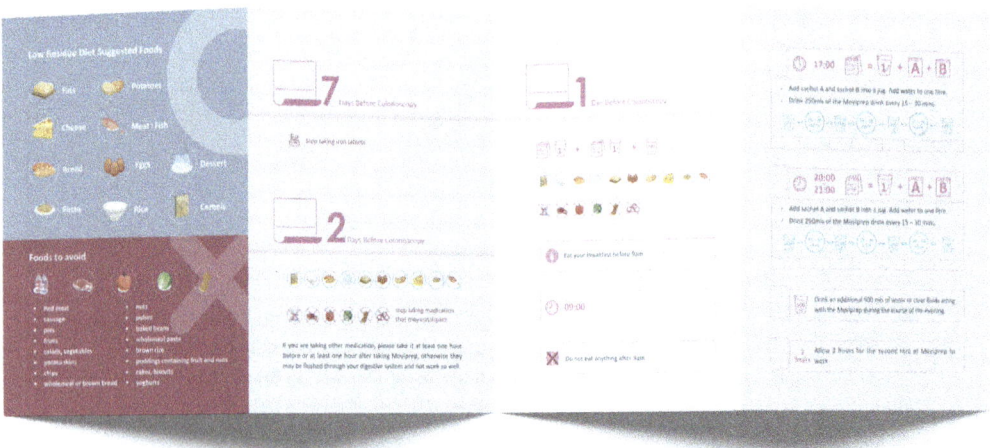

Figure 116 Final detachable schedule

Figure 117 Final redesigned booklet mock-ups

Figure 118   Final redesigned booklet (all pages)

Figure 119   Final redesigned booklet mock-up

MOTION GRAPHICS

As patient-support motion graphics was an alternative and/or companion version of the booklet, it followed the same design style and colour palette as the redesigned booklet. Creating a second output from an output that had been already tested saved time, i.e., less iterations, and gave the research team better insight regarding the design. For example, insight on: what information needed to be communicated more effectively than in the original booklet; what worked well in the new booklet and should be kept; and what was problematic and should be avoided. In terms of problems, things to be avoided were: lack of clarity regarding foods allowed and foods prohibited; lack of clarity on how exactly to take Moviprep purgative solution, etc. The motion graphics design was also based on systematic literature reviews on information and motion graphics design.

There were several stages of design development and iteration where representatives of the target audience, as well as information design experts, were asked to watch the motion graphics and provide feedback. Results were as follows:

- Layout – improve organisation of information;
- Transitions – make transitions smoother;
- Video length – reduce length to no more than 4 or 5 minutes, otherwise it is too tiresome;
- Video pace – speed up the pace to make the video more dynamic and shorter;
- Action – give movement to some objects to make the video more active;
- Key words – make key words bigger;
- Sound effects – adjust to make sure sound effects fit with what is shown on the screen and do not over use;
- Voice over – redo to make more engaging;
- Voice over script – explain particular information more clearly, such as prohibited and allowed foods and how to take purgative solution;
- Background music – change to a more balanced tone, i.e., not too cheerful, but not too serious either;
- Content and respective illustrations – adjust the content for clarity, i.e., add more specific information on diet list, such as 'white bread' and not just 'bread', emphasise 'no potato skins', and clarify medicines using plain English, such as using the word 'iron' instead of 'Fe', etc.

Four participants were then tested using the same template of questions as the one to be used for the experimental testing in order to fine-tune the final design further. The final design of the motion graphics is shown through static frames on the next page.

Final motion graphics

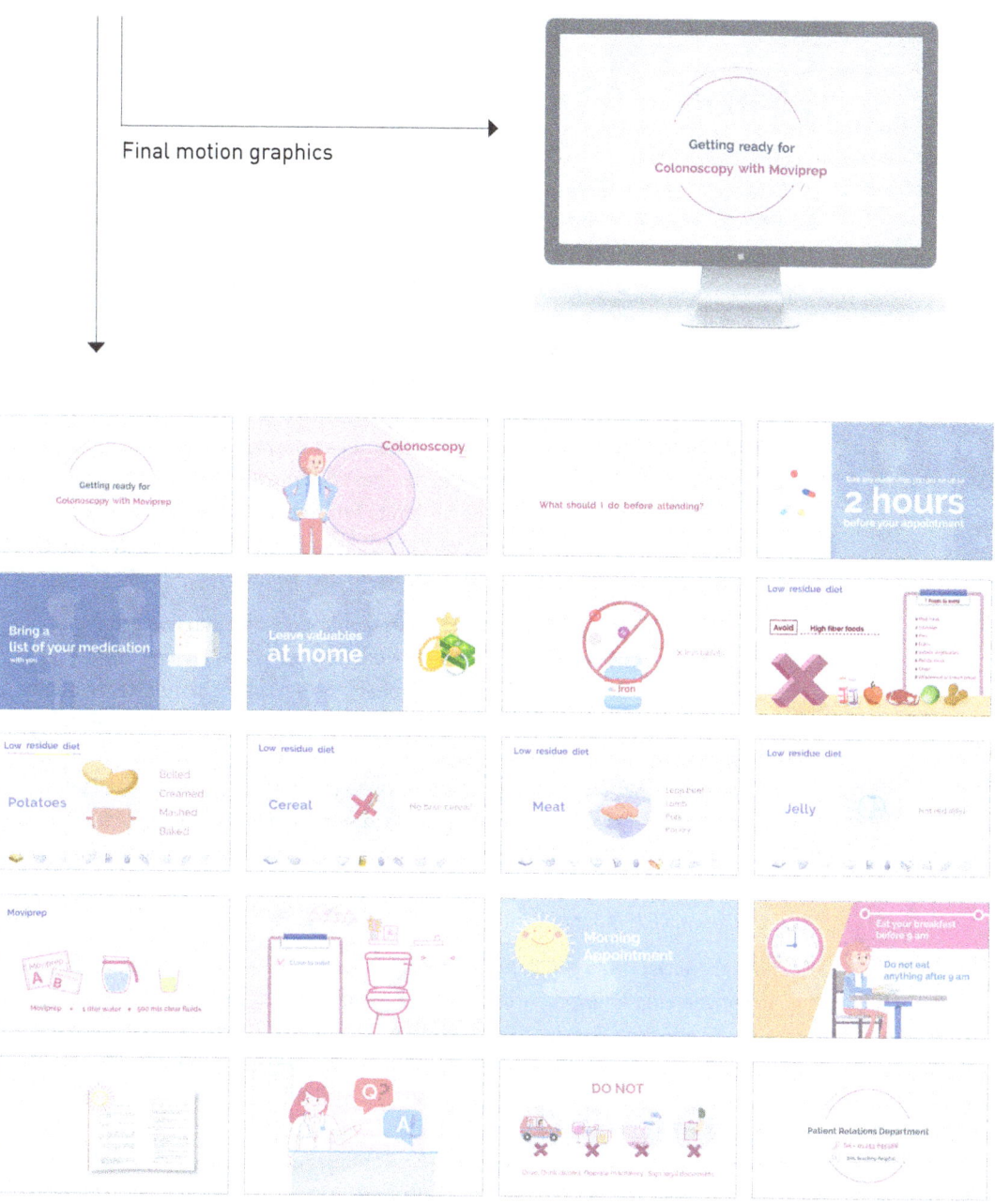

Figure 120   Static frames from the final motion graphic output

USER PERFORMANCE TESTING

Two stages of experimental testing and statistical analysis were conducted to measure participants' performance (i.e., the time to find information and comprehension accuracy of the information found): Stage 1 tested the booklet and Stage 2 tested the motion graphics. Each stage tested a sample of 60 participants who had not undergone a colonoscopy before or had no knowledge of how to prepare for a colonoscopy (e.g., if caring for a family member who had to do a colonoscopy). It was important that participants did not have knowledge of bowel preparation information, otherwise it would mask the results as they could guess the answers without having to find it in the booklet. Each participant was tested individually and given a task to complete, which asked them to find specific information in the booklet or motion graphics, to answer questions about bowel preparation for colonoscopy. Participants were then asked their opinion about the design of the booklet.

In Stage 1 (booklet testing) participants were equally divided into two groups of 30 participants each: Group 1 used the old booklet design and group 2 used the redesigned booklet. In Stage 2 (motion graphics testing) participants were equally divided into two groups of 30 participants each: Group 3 included Gen X users (i.e., an older generation) and Group 4 included Gen Z users (i.e., a younger generation). The reason for testing two different generations was to ascertain whether the claim that younger audiences are more used to motion, are more attracted to it, and perceive it more easily and naturally (Strizver, 2014) is supported in the context of preparing for a medical exam, or whether motion graphics is superior to static graphics (Höffler and Leutner, 2007) independently of the audience using it.

Statistical analysis was conducted to compare performance for completing the given tasks between the following groups: G1 – Existing Booklet Design; G2 – Redesigned booklet; G3 – Motion Graphics Generation X; G4 – Motion Graphics Generation Z. The results were very clear in showing evidence (a statistically significant difference) for the superiority of a Redesigned Booklet developed following research-based information design principles and user-centred research methods, in communicating bowel preparation information effectively. This superiority is even higher with the Motion Graphics following the same principles and methods.

The results also show that the Motion Graphics is as effective in communicating information to a younger generation (Gen Z) as it is to an older generation (Gen X).

Participants were then asked their opinion about the design of the output they used. The vast majority of participants agreed that in the Redesigned Booklet the information was: easy to find; easy to understand; memorable; engaging; and trustworthy. The comments made by participants were in line with information design principles as they agreed with the benefit of, for example:

- Using large text for headings and key points, and using colour and bold for emphasis, as it makes it easy to find the information;
- Having the information divided in different sections, as well as in a sequential order;
- Having diagrams and step-by-step instructions to help understand what to do;
- Having icons and illustrations that make the information easy to understand, as well as engaging and memorable.

The vast majority of participants also agreed that the Motion Graphics was: clear, easy to follow, and that the design style, colour palette and typography were appropriate. Participants also agreed with the benefit of, for example: having a good illustration style and calming colour palette; paying attention to detail and giving an overall feel of a professional and well-made motion graphics; putting people at ease before doing a colonoscopy, etc.

CONCLUSION

In this project, a user-centred multiple-methods approach was followed in three sequential stages that mirror the stages of my research and design process model as proposed in Chapter 4: (1) Discovery – survey of booklets, focus group, online questionnaire; (2) Development – usability testing and iteration; (3) Evaluation – experimental comparison between four groups and three different outputs.

The results showed the superiority of information combining text and visualisation over text-dense information. The motion graphics, in particular, combining written, visualised and spoken information, enabled explanation in more detailed step-by-step instructions, as well as illustrating exactly what was required in a much shorter period of time than reading several pages of text.

In conclusion, this study has shown alternatives to face-to-face support, as the latter can be time consuming, costly and it is not always accessible to patients. On the one hand, the effective combination of text and visualised information (following research-based design principles and applying user-centred research methods) can help colonoscopy patients find information more easily and improve comprehension of complex information for bowel preparation for screening. On the other hand, having at patients' disposal various multimedia outputs (booklet – print and pdf – and motion graphics) will make bowel preparation instructions more inclusive and more accessible to a larger audience. The more effective and inclusive health information is, the easier it should be for patients to manage their health. Consequently, it could increase uptake and compliance, improve quality of bowel preparation, reduce anxiety in patients, increase colonoscopy completion and the accuracy of diagnosis, as well as reduce colonoscopy time, patient pain and discomfort, and costs associated with longer and failed colonoscopies.

5.2 | CASE STUDY 2 | VIRTUAL LEARNING